Quantum Energetics
and Spirituality

Quantum Energetics and Spirituality

Aligning with Universal Consciousness

Volume 5

KENNETH SCHMITT

Copyright © 2022 by Kenneth Schmitt

All rights reserved. No part of this book may be reproduced or transmitted in any form or by any means, electronic or mechanical, including photocopying, recording, or by any information storage and retrieval system, without permission in writing from the author.

ISBN: 979-8-9851064-4-2

Kenneth Schmitt
Phone: **+1-808-280-4041**
Email: **Ken@ConsciousExpansion.org**
Website: **https://www.ConsciousExpansion.org**

Independently published by the author

Contents

Introduction ... 1

   Why We Must Become Fearless ... 3

   Living in a Parallel Reality ... 7

   A Path of Transformation ... 9

   Aligning with Ascension Energies ... 11

   Becoming a Spiritual Super Athlete ... 13

   Being the Persons We Really Want to Be ... 15

   A Path of Personal Transformation ... 17

   Choosing the Dimensions of Our Experiences ... 19

   Living in Intense Situations ... 21

   Transformation is Our Choice ... 23

   Transitioning through the Precession of the Equinoxes ... 25

   The Transformation of Humanity ... 27

   Making a Leap in Consciousness ... 29

   Living in Our Increasingly Intense Energetic Environment ... 31

   Creating the Life We Truly Want ... 33

   Realizing Our Eternal Self ... 35

   Transforming Chaos and Hatred into Compassion and Joy ... 37

   Following Our Inner Light ... 39

Creating Our New Reality	41
Increasing Our Inner Radiance	43
Opening Ourselves to Greater Awareness	45
Aligning with Our Inner Self	47
The Magic that Yields Ultimate Freedom	49
An Enlightened Life	51
Our Identity beyond Ego-Consciousness	53
Resonating with Heart-Consciousness	55
A Path to Personal Transformation	57
Experiencing Levels of Realization	59
Working with Our Energetic Alignments	61
Identifying Our Essential Self	63
Living Beyond Fear	65
The Evolution of Human Consciousness	67
Experiencing the Vibrations We Love	69
Realizing Our Personal Truth	71
Creating a Fulfilling Life in a Time of Great Challenges	73
Understanding and Transcending Our Limitations	75
Awareness Beyond Time/Space	77
Facing and Resolving our Personal Challenges	79
Reaching for a Deeper Understanding of Life	81
Our Inward Journey to Infinity	83
Transforming Ourselves in Our Daily Lives	85
Higher-Dimensional Living	87
The Extent of Our Consciousness	89
Deepening Our Realization	91

Living as Our True Selves	93
Aligning with Our Cosmic Environment	95
Transitioning with Earth-Consciousness	97
Conscious Human Evolution	99
Choosing Our Quality of Life	101
Insights into the Nature of Our Reality	103
Life as a Play of Consciousness	105
Expanding Our Realization	107
Creating a Presence of Love	109
Knowing beyond Our Human Ability	111
Expanding Awareness of Greater Consciousness	113
Mastering Personal Transcendence	115
Our Participation in Universal Consciousness	117
Aligning with our Essence	119
Realizing Unconditional Love	123
Being in Time and Out of Time	125
Awareness of Greater Consciousness	127
Aligning with Our Greater Self	129
Living in the Flow of Life-Enhancing Energies	131
Creating a Viable Perspective	133
Creating a Wonderful Life	135
Understanding Our Potential as Creators	137
Our School of Enlightenment	139
Realizing a New Dimension of Living	141
Removing Fear from Our Lives	143
Our Higher-Dimensional Essence	145

Opening Our Awareness to Personal Regeneration	147
Realizing Our Inner Light	149
When Thinking Arises from Knowing	151
Opening to Infinite Awareness	153
Living an Inspired Life	155
Realizing Our Creative Essence	157
Recognizing and Transcending Our Limitations	159
Opening to Our Transcendence	161
Anticipating Life with Confidence	163
Opening To Expanded Awareness	165
Expanding our Realizations	167
Transcending Our Empirical Trance	169
"Be still and know that I AM God"	171
Realizing Infinite Love and Joy	173
The Origin and Resolution of Personal Limitation	175
Conscious Light Is Enveloping Our Awareness	177
Examining the World Within Our Consciousness	179
Becoming Absolutely Free and Sovereign	181
Adapting to Our Changing Cosmic Environment	183
Being the Persons We Love the Most	185
Rising into Our Magnificence	187
Expanding into Greater Realization	189
Transforming the World of Our Experience	191
Observations on Realizing a Greater Reality	193
Standing Powerfully in the Face of Fear	195
Experiencing a State of Knowing	197

Aligning with Our Conscious Essence	199
Training Ourselves for Greater Awareness	201
Living Beyond Our Limitations	203
Realizing the Mystery of Life	205
Transcending Ego-Consciousness	207
Recognizing, Resolving and Releasing Our Limiting Attachments	209
Developing Heart-Conscious Awareness	211
Expanding Our Awareness beyond the Ego	213
Accepting and Adjusting to Our Ascension	215
Transforming Humanity	217
Living with Conscious Intention	219
Realizing Our Divinity	221
Living in a Dimensional Shift	223
Awareness of Inner Guidance	225
Living beyond Our Current Reality	227
Moving from Unintentional to Intentional Creation	229
A Possible Understanding of Crystalline Time	231
Finding Ourselves Within	233
Achieving True Personal Freedom	235
Our Transition to a New World	237
The Function of Gratitude in Our Lives	239
Realizing Our Divine Nature	241
The Magnificence of Quantum Healing	243
Learning to Direct Our Lives Intentionally	245
Beyond the Trance of Humanity	247
Awareness of Inner Knowing	249

Returning from Our Sojourn in Duality	251
Expanding an Understanding of Life	253
Some Insights into Our Game of Life	255
Living Beyond Duality	257
Our Great Opportunity as Humans	259
The Gathering of Our Spiritual Family	261
Developing Inner Knowing	263
Transcending Fear and Limitation	265
Realizing Our Infinite Self	267
Self-Transcendence	269
Choosing Self-Suppression or Enhancement	271
Human Life as a Game of Consciousness	273
Understanding Artificial Intelligence	275
Realizing Personal Mastery	277
Transforming and Elevating Our Lives	279
Insights into Our Humanness	281
Becoming Aware of Our True Essence	283
Realizing Complete Freedom	285
Enhancing Our Lives	287
On the Way to Infinite Awareness	289
Transforming Ourselves and Our World	291
Living in the Creator's Life-Stream	293
Exploring beyond Ego-Consciousness	295
Mastering Empirical Reality	297
The Nature of Our Divinity Within	299
Dealing with Negative Situations	301

The Next Evolution of Time/Space	303
Overcoming Survival-Consciousness	305
Realizing the Greatness of Who We Are	307
Exploring Our Belief in Good and Evil	309
Creating a Life of Beauty and Majesty	311
Our Deepest Love	313
Examining Our Core Consciousness	315
Accepting Our Invitation to Conscious Expansion	317
Insights into Universal Consciousness	319
Shifting beyond Ego-Mind into the Unknown Void	321
Choosing and Experiencing Our Quality of Life	323
We Are Energetic Transformers	325
Directing the Living Play of Energies	327
Intentionally Changing Dimensions	329
Insights into Our Expansive Potential	331
Processing Our Limitations	333
Understanding and Expanding Our Humanity	335
Acknowledging Our Inner Light	337
Empowering Our Transformation	339
More Fully Developing Our Capabilities	341
Living Ineffably While Being Incarnated	343
Working with the Energies of Relationships	345
Transforming Aging to Ageless	347
Realizing the Truth of Who We Are	349
Navigating the Split in Reality	351
Living in the New World	353

Achieving Transcendence	355
Creating the Life We Desire	357
Directing Our Mental and Emotional Alignment	359
Going Deeper into Self-Realization	361
Using Our Emotions for Transformation	363
Transforming Time and Space	365
Realizing the Importance of Our Attention	367
Entering a State of Transcendence	369
Disempowering Evil Control of Humanity	371
Realizing Infinite Awareness	373
Compassion and Love Are Rising	375
Realizing the Essence of Our Consciousness	377
The Value of Pain and Suffering	379
Living Intentionally in the Field of Consciousness	381
Our Destiny and Our Choice	383
Examining and Enhancing Our Inner Life	385
Imagining and Realizing Our Reality	387
Deepening Our Self-Understanding	389
Impressions of Our Expanding Consciousness	391
The Challenge of Physical Death	393
Dealing with Tyranny and Adversity	395
Transforming Our Consciousness	397
From Situational to Sensational Awareness	399
Deepening Our Understanding and Expanding Our Awareness	401

Introduction

My books are essays that arise in my intuitive knowing. I write about my inner experiences, and the process that I go through in being able to realize expanded consciousness. I write in the plural, because we are all the same consciousness playing the human drama with our personal experiences. Realizations come to me, penetrating deeper into my limiting beliefs, and I write about the process I go through to enter greater awareness.

Many of my realizations are based on quantum physics and are logical extensions of physicists' findings in the existence and pervasiveness of consciousness. Ancient esoteric spiritual teachings also prompt my deeper understanding. What I most love is writing about these things at this time of transition for humanity into a brighter expression of life. The cosmos is inviting and propelling us into expanding awareness of Self-Realization.

Born with innate knowing of our essence, but without a clue of how to live in the energetically dense, largely negatively-polarized world of humanity, we telepathically pick up the perspectives and beliefs of our family and society, enclosing our awareness within the limits of human belief systems. We develop our ego-consciousness to enforce these limitations, so that we can have an authentic human experience without higher guidance. But no one beyond our own consciousness enforces our limitations. Innately we are completely free with infinite awareness and unlimited creative power. We are fractalized extensions of universal consciousness.

Once we realize the truth of who we are, we can no longer be victims. We become masters of our lives. We can understand the play of consciousness that we are involved in. As we direct our state of being and act out our destined roles, we create the qualities of our experiences by the way we are, primarily how we feel and what we believe about ourselves.

Realizing the dualistic, empirical world as our reality, and believing that nothing exists beyond our empirical experience, prevents us from awareness of a greater reality. If we want to open our awareness to a greater understanding and mastery of life and knowing our essence beyond time and space, we must pay attention to the consciousness of the source of our life-force. This helps us transcend dualistic limitations and open to infinite consciousness. These are the impressions that I discuss in this book. The essays are presented in the order that they came to be. There are no chapters. It is all a chapter in life.

Sending you deeper insight, joy and fulfillment in every way,

Kenneth Schmitt
September 15, 2023

Why We Must Become Fearless

To be the best and most potent versions of ourselves, we must become fearless. We can transform the contracting energy of fear into the expansiveness of joy by our personal choices. We can learn to use our free will creatively and without limitation. We have learned to choose to live in doubt and fear, and we have made conditions that express the vibratory level of doubt and lovelessness. This is not a natural limitation. We have created it, and we can transform it. For us, it is a conscious choice. Once our sub-conscious is imprinted with our choice, regardless of why we made the choice, it establishes the polarity and vibratory level for our state of being, bringing our thoughts and emotions into alignment with its energetics. As long as we harbor doubt of any kind, we are experientially enslaved to its polarity, which is negative.

We can choose love only by being fearless. When we feel doubt about anything arise in us, it is a signal for us to pay attention to what is going on. Instead of letting our ego-consciousness direct us into a situation in which we compromise our intention to be grateful, compassionate, loving and joyful, we have the power of choosing our vibratory level. In any moment, we can choose to be however we want mentally and emotionally, regardless of how imposing conflicting energies may be.

In realizing that the power of our free will is an important ability, our intuition is our assistant for higher guidance. This is our heart-felt inner knowing, and it is constantly flowing wisdom and guidance for us to maintain the most positive perspective we can allow for ourselves. We all have it. We just need to pay attention to it and choose to follow its guidance. It is what we really know. The truth is self-evident in how we feel about it. It's always a choice between what is life-enhancing for all, and what is life-diminishing.

If we decide to be positive, we cannot be negative at the same time. This means that we either participate in the world of duality that is the human experience, or we transform ourselves by creating vibrations that result in our physical, mental and emotional processes, as well as our environment, becoming uplifting in every way. Because of our perspective, based in gratitude, joy, compassion, and love, we can have wonderful lives in every moment. Essentially, we can live in a parallel density, but in a higher dimension.

Negativity can exist only if we create it within ourselves, which we have done for eons. We have believed that it is inherent to the reality that we recognize. It is. Duality requires both positive and negative. It could not exist without the light, because light is the source of existence. It is our conscious life force. The negative consumes light and blocks its radiance, converting its energy into self-destructiveness. All of this may happen In subtle and even confusing ways for us, because it is different from the way we have learned to expect to experience life.

We have been fearful all the time, creating fearful experiences for ourselves. To be fearless is to know and trust our intuition absolutely. By following what we innately know, that stimulates us to feel gratitude and joy, we can trust that we're aligned with the highest vibratory level. This is where the desires of our heart become clear. By following them, we enhance our lives in every way that we allow. With absolute confidence we can choose the energetics that we want to express within ourselves and radiate into the quantum field for manifestation among humanity.

Living in a Parallel Reality

If we want to experience living with the most wonderful feelings and visions, we can open ourselves to the highest positive vibrations that we are capable of receiving and sharing through our heart. In our creator essence, we receive awareness of energetic patterns that we open ourselves to. These energies we can modulate into our personal vibrations in alignment with the enhancement of all life. Every scenario in our experience has this potential. Our experiences are created by our choices of energetic alignment in our thoughts and emotions. Divine magic occurs when we choose to live in the dimension of gratitude and joy in every moment.

We can enter an energetic world that is the same as the dualistic energies experienced by most of humanity, but our personal experiences can have the quality of strong expressions of our heart-consciousness. We do not suffer in the realm of duality, even though we are aware of it. When we intentionally create our personal reality through our realization of it, our lives can be filled with wonderful experiences.

Since everything in our lives is an energetic expression of consciousness, our world arises for us from our realization of it. With our open imagination, and with clarity in our emotions, we can create our energetic state of being, expressing the radiance of energetic formations through our full consciousness. As we learn to penetrate the hidden, self-imposed, energetic patterns in our subconscious, we can fulfill our deepest needs for acceptance, love and absolute support in every way. We can intentionally open our awareness to our heart-consciousness by aligning with its vibratory resonance in gratitude, joy, compassion and love.

The hidden fears, doubts and all negative energetics in our subconscious self are within us, because we have become attached to them, often through trauma and our family energetic inheritance. They live off of our life force, and are waiting for us to accept them and thank them for allowing us to have experiences that are impossible for us in our true infinite awareness. As a result of negative experiences, we are wiser and have greater compassion. Once we realize this, we no longer need these experiences and can release them in realizing our infinite Self in our eternal presence of awareness.

By intentionally thinking about and feeling life-enhancing scenarios, and transforming negativity into positivity through our imagination and emotions by intentionally resonating in alignment with our heart-consciousness, we can transform our lives into experiences that we are truly grateful for in every way, and that manifest joy, love, abundance and freedom in ourselves and all around us.

A Path of Transformation

When we're enveloped in the wilds of nature, perhaps high in the mountains with panoramic views and silence all around us, and feeling great gratitude and joy, we can ask the angels of the air to brighten our experience even more, which they love to do, when we ask and are open and receptive in complete confidence, because of our resonance with the nature spirits, the Spirit of the Earth and our heart-consciousness. When we are in alignment with our heart-consciousness, we can communicate telepathically with all beings and with the consciousness of all things.

When we align with our heart-consciousness, we can be expansive and ecstatic in the thorough enjoyment of being who we are in the magic of our powers of awareness and creative expression. This is our natural state of Being and is our experience, when we resolve and release our limited beliefs about ourselves. We can then enter the vibratory spectrum of absolute confidence in the creative impressions and expressions of our focus of attention. By aligning our perspective on life with the vibrations of joy and compassion, we can deeply understand every moment and live in our connection with universal consciousness and enhancement of all life.

By opening ourselves and wanting the awareness of our heart-consciousness, we can leave the world of duality and enter a higher dimension of expanded conscious awareness in unconditional love, personal fulfillment and infinite creative ability with our mental and emotional expressions. We can learn to use the power of our imagination to receive the expressions of heart-consciousness. We can call up emotions that feel the qualities of the vibrations that we are paying attention to, and we can be present as much as we desire at this vibratory level.

Living in the realm of gratitude and joy, we can be exempt from all negative experiences, because we realize that in the vibratory spectrum of heart-consciousness, there is always quantum support for us to achieve what our heart desires. The quantum field manifests our experiences in resonance with our predominant thoughts and feelings. As we learn to direct our thoughts and emotions to life-enhancing visions and scenarios, as well as just being present in gratitude and love, we leave the realm of duality and enter the realm of life experiences that become magical and wonderful. In this way we can transform our lives, and it can happen whenever we choose with strong enough intent of our entire consciousness.

Aligning with Ascension Energies

Our world of dualistic reality is being transformed into a higher, more positive vibratory band of energetics. Dualism is dissolving. Our world has been freed from negative control, and the energetic environment that we share is becoming a shared consciousness of the awareness of everyone. We can telepathically realize our natural state of Being as luminous Beings of eternal presence of awareness, endowed with infinitely powerful creative ability. We are fractals of Creator consciousness, awakening from of the hypnotic trance of humanity.

By paying attention to our intuitive knowing and aligning our feelings with the energy of our heart. In a very expansive way, we can elevate our own vibrations into joy and compassion. We can merge into our subconscious and resolve our limiting beliefs about ourselves. As we intentionally raise our vibrations positively, fear and doubt transmute into another dimension. We are now stimulated only by our own conscious awareness. We no longer need to stay in the compartment of consciousness that we have lived in.

As we resolve our limiting beliefs about ourselves, our ego-consciousness disappears, because it has been based on the negative vibrations of fear and doubt. Aligning with negative polarity, we could not be aware of our heart-consciousness and higher guidance. It is now easy to release our attachment to the limitations we accepted in our subjection to strong negativity, which no longer exists on this planet. What still appears to be negative governing power is only a show. It has no essence any more and can control us only if we believe it's real and engage with its energetics.

Instead of reacting in life-diminishing ways to negative situations in our lives, we have the choice of paying attention to our heart-consciousness and being in gratitude and joy, even in ego-threatening situations. We can understand the energetics that are present and can transform them in our imagination and emotions by intentionally aligning with our heart-consciousness in awareness within universal consciousness. Because this vibratory level of awareness is in a different dimension from negativity, we can choose to be brought into alignment only by positivity. This is life-transforming for us. Our experiences manifest the energy of our heart, bringing us expanding joy and fulfillment.

Becoming a Spiritual Super Athlete

Super athletes are able to keep being more daring and to keep pushing human limits by convincingly imagining themselves doing it. This is also how we can achieve expanded consciousness and mastery of the material world. It is available to all of us. We just have to convincingly imagine and feel ourselves living in scenarios at its vibratory level. This is a challenge for us, because of the depth of programming that we have accepted about ourselves as limited in our ability to realize our true nature and infinite presence.

Imagining ourselves doing something extraordinary beyond physical feats and even psychological feats, is part of our creative ability. As we express our thoughts and emotions, their energetics radiate within and around us, attracting compatible energy patterns to provide experiences for us. When we are confident and assured, we modulate the energy patterns around us and within our awareness into resonance with us. Apart from any other influences or circumstances that we may be participating in, we have the ability to direct our thoughts and emotions through our imagination into alignment with our natural vibratory level in infinite presence of awareness.

Achieving infinite awareness comes naturally for us, once we free ourselves from believing that we are incapable of it. We have to resolve our limiting beliefs about ourselves and move beyond them. This is what makes a super athlete. We can all do this, once we can convince ourselves that we really are limitless.

A good practice can be examining our personal beliefs. What is their essence, what are they based on? We'll find that all limitation is based on fear of some kind. It is produced in our own emotions. It does not exist outside of us. Because of the belief in suffering and mortality, fear is a necessary component of the human experience. The clue here is mortality. For a person enclosed within the conscious limitations of duality, it can be helpful to read accounts and watch videos of people who have had convincing out-of-body experiences and have returned to their momentarily dead bodies to tell us about what happened.

Our conscious awareness does not die; instead, we become much more expansively aware and without personal needs to satisfy. This is our natural state of being. In our essence, we are our presence of awareness. We have absolute control of our own energies always, whether expressing ourselves through our physical bodies or beyond them, but we must learn how to direct our attention and how to align emotionally in every moment with the energetic levels that are enhancing for all life, and not have interference from our ego-consciousness.

Being the Persons We Really Want to Be

Within the realm of our participation in duality, the negative force became strong enough to be much greater than the dimension that we recognize. Negative-creating beings learned how to enter our consciousness and find vulnerabilities that could cause us to compromise our intentions. These are vibratory levels that we experience when we assimilate the belief that our consciousness will ultimately be terminated. This is the fundamental belief that is the basis for all of our other limiting beliefs about ourselves. This is the basis of all of our fears, and we must resolve and transform it into the experience of our true essence as our infinitely-powerful-creator-presence of eternal awareness.

It is impossible for us to be terminated. In the world of empirical duality, our awareness is limited to one dimension of density in consciousness, but our physical presence is only a partial expression of ourselves. We know from the reports given by those who have died physically, but then came back, that our awareness expands greatly upon disembodiment and unattachment from limitations. Without the body we are our pure essence of conscious awareness, unlimited in every way and able to express ourselves in any way that we imagine. We can know everything about everyone and everything. Our consciousness envelops the cosmos and beyond, and universal consciousness is our realization.

Once we learn to release attachments to limitations, we are truly free to be the persons we deeply long to be. We must face our belief in personal termination. We can know that we exist beyond all energetics, and we express our vibratory presence energetically. In our presence of awareness beyond time/space, we have free access to the consciousness of the Creator and the awareness of all conscious creatures. At this point, we participate differently in the world of human experience.

We can be aware in every moment that any presence of negativity in our personal experience is a result of our mental and emotional recognition, alignment and engagement with it. When we do not give it reality through our realization, it cannot exist for us. Every scenario in our experience has an infinite number of ways for us to experience it, and we are free to choose any of them. Without fear and doubt, we are free to create all the desires of our heart for our experience.

When we can live in the realm of gratitude, love and compassion, we can have confidence in our absolute creative ability just by our state of being. We create our personal energy signature, which then attracts and repels other energetic patterns. The ones that resonate with ours provide our experiences. Any of us can become masters of our lives, and we can experience the most wonderful relationships in deepest love and complete sovereignty.

A Path of Personal Transformation

As humans living in the spectrum of vibrations that we realize as real, we have a great challenge to realizing a more wonderful reality. Once we decide that we want to elevate our energetic expression, and live in a realm of love and joy, we have many options to expand our realization beyond our current limitations. If we feel imposed upon by a great force of negativity that threatens our physical lives, we may want to dissolve this experience, but we have trained ourselves to realize its reality, whenever its vibrations grab our attention. We feel compelled to engage with it. This brings our own thoughts and emotions in alignment with it, either for compliance and support or for resistance and anger. Both of these perspectives are the same kind of energy. It is restrictive and diminishing of life.

Until we are able to master the dualistic empirical world, we are governed by our ego-consciousness, which must protect itself from its believed suffering and mortality. This happens in us when we focus on the negativity of fear and doubt. We feel that we have no choice, and that we must deal with negativity on its own level. The more we engage with it, the more we give it control of our awareness. Our ego-consciousness feels that we must suffer in some way in order to survive and live comfortable lives. When we have no awareness of higher guidance, this is a true situation.

If we want awareness of our true intuition and divine guidance, there is a shift in our awareness that we must initiate intentionally. If we want expanded awareness, we must align our own vibratory level with gratitude, compassion and joy. We can recognize the quality of energies that we face, while maintaining our inner awareness as well. This is how we can realize and align with our inner knowing. Intuitively, we can participate in the consciousness of the Creator, and we can be unlimited in every way. When we can continue to feel unconditional love for the consciousness of everyone, regardless of their state of being, we can live in miraculous ways, while resolving and releasing our own negativity, which transforms our lives into experiences of gratitude and joy. We receive the feedback from our own vibratory level of love.

When we can stay completely positive in gratitude and love in every moment, while being in alignment with the vibrations of our heart-consciousness, we are free in every way, and we can choose to create without limit the experiences that resonate with our heart's desires. This is how we transform our lives when we intentionally use our imagination and emotions to create experiences in alignment with our realization of the reality of love and light in the conscious life force of everyone. We can become master modulators of all patterns of energy by the power of our realization.

Choosing the Dimensions of Our Experiences

The quantum field of all potentialities contains the vibrations of every experience for us to have in our lives In any moment. Our destiny is to live in positive, high vibrations that we realize are real. These vibrations are life-enhancing and enriching in every way. They bring us closer to being able to realize unconditional love in our connection with all conscious beings. Every entity from the smallest sub-atomic swirling essence of life, to the universe itself, exists within our own consciousness. We participate in the consciousness that creates everyone and everything.

Our personal intuition comes from universal consciousness through the consciousness of our higher chakras, primarily our heart. When we imagine and feel their vibratory range, we can align with them, but when we are misaligned, we cannot know our higher guidance. Our ego-consciousness then controls us from its limitations based on fear of suffering and survival. These are imaginary.

Always we experience the best situations that we allow for ourselves, along with course corrections directed by our unseen guides, who know what we truly need in order to help us to be aware of our purpose and goal in this life. We can be aware of the symbolic meaning of everything in our experiences. We can awaken from the hypnotic trance of humanity that is focused on dualistic empiricism. When we choose to entertain any kind of fear or doubt, we align ourselves with negativity and cannot create positive experiences. Duality is a range of energies that allow for the focus on negativity. If we want to live in a realm of love and joy, we must be aware of these energies and feel them within ourselves.

In our essential limitless presence of awareness, we can easily resolve and transcend our limitations through acceptance of our experiences and gratitude for what happened to us as a result. When we haven't taken the small prompts that our guides give us, they resort to more powerful guidance. We are the ones who make all the decisions about the energy that we align with and experience. When something major prevents us from living the way we have, it's an opportunity for us to reevaluate our life path and open ourselves to our inner guidance.

We live within the energetic spectrum that we resonate with, and while we continue to choose to allow negativity to control our awareness, we cannot be aware of the joy of the higher dimension that we coexist with. Once we consistently and intentionally choose to feel and imagine experiencing what we truly want the most in our deepest consciousness, we open our awareness to the vastness of our unlimited essence. We become aware of our entrancement in dualistic empiricism, while opening our realization to a greater spectrum of energies and experiences.

Living in Intense Situations

When we are in intensely magical situations, and sometimes very dangerous ones as well, we are released from our normal human trance. We are brought face to face with the belief in our mortality. It's exciting and terrifying at the same time, and we can choose which it is to be. We have an opportunity to choose to vibrate in resonance with the fear of death or with the joy of life.

In intense situations, we are filled with compassion and gratitude as we help one another. In these situations we are able to transcend our limitations and to follow our heart-consciousness. We may feel a deep bond with one another to have transcended ordinary life in the face of imminent death. When we're ready to drop our attachments to personally-limiting beliefs, we may experience this kind of symbolism to jolt us into a greater realization of who we are as Self-Realized, eternal presence of awareness with unlimited abilities. Our ego-consciousness cannot imagine this, and words cannot convey our greatness. Mastery of our lives awaits our realization, and intense experiences direct our attention toward transcendence.

When once we have an experience of transcendence beyond ego-consciousness, we know what it is, and we can align with its positive, high vibrations, while opening our awareness to it. There is no requirement for us ever to leave our Self-Realization, and we can continue to live our human life in a profoundly different spectrum of energetics, in which life is enjoyable and easy. Our environment is designed to support and enhance life.

When we are in alignment with these energies, we are provided with everyone and everything we need and love. The energies of life-enhancement radiate from our physical and psychic presence.

Situations arrange themselves in our personal lives to bring us experiences that we resonate with. Our physical bodies also collaborate with our personal energetics, resulting in either life-enhancement or life-diminishment. Our personal choices determine which kind of energy we experience. Even our DNA can awaken in response to the stimulation of greater light in our awareness. Every aspect of life offers us an opportunity to contribute to its enhancement by our conscious connection. All we need is a high state of being in transcendent awareness. Everything around us comes into resonance with us. Instead of submitting to limitations and negativity, we can choose to realize a higher dimension of love, compassion and joy. Whatever quality of energy we pay attention to and recognize feeds our realization of what is real in our experience.

Transformation is Our Choice

When, we pay attention to the highest feelings that we can imagine in our heart-consciousness, we can feel the greatest love and joy. At this level of vibrations, we are beyond negativity and any kind of fear or doubt. How much intensity of love we receive depends upon our ability to stay open and receptive to it. This is an intentional journey that we can pause at any time and reorient ourselves. We can be in the process of taking on a completely positive perspective, knowing that we are cared-for by universal consciousness, which wants to give us everything we love. We can open our awareness to this knowing, resulting in the transformation of our lives as much as we allow in every moment.

By living in positive, high-vibrations of compassion, joy, and gratitude, we can realize our presence beyond the physical world. The dualistic energetic spectrum that humanity inhabits consciously has no power over us, unless we align with it. By paying attention to the energetics of our heart-consciousness, we can open our awareness to our eternal presence of Self. We live in a consciousness that is beyond energetics, and it commands them. Even now, even if we don't understand or know about this, we are continuously creating experiences for ourselves by forming vibratory patterns of electromagnetic waves as expressions of our thoughts, emotions and of the realizations in our consciousness.

As we transform our lives, we learn that we can modulate the energetics of every situation that we participate in. By following our inner knowing in every moment, we enter a dimension of love and beauty. This is our natural state of being, and we are being drawn into it by the increasingly powerful environment of Light, in alignment with the energy of our heart-consciousness. It encompasses the Earth and beyond, even beyond time/space. It is universal consciousness, and we participate in as much of it as we can open our awareness to. This is part of our intentional alignment with higher-dimensional energetics in greater love and ecstasy.

We can become intentional in every moment, and aware of our inner guidance. By releasing our attachments to limitation and fear, we transform into transcendent Beings. Even our DNA changes on an etheric level and then manifests in our experience. As magic and miracles become normal experiences in our lives, in alignment with our energetic expressions, ego-consciousness becomes our servant, rather than the master it has been. Our heart-consciousness shines brightly in our awareness and opens areas of consciousness that we had closed off from ourselves. As we go about our lives, everything changes for the better, and our radiance elevates the consciousness of humanity.

Transitioning through the Precession of the Equinoxes

As we transition from one 25,920-year era to the next, we are facing a shift in consciousness that is more massive than almost any of us can imagine. We have been thoroughly trained to accept and even embrace negativity in every area of life. Facing life without negative control is cause for great celebration and rejoicing, but how can we know that we have been freed from negative control, especially when most of the media continues to announce even greater limitations to our freedom? How can we know that those limitations and the tyranny that they arise out of are no longer powerful?

As with transcending all limitations, freedom comes through realizing the energetic influence of our heart-consciousness. This is the only stimulation in our world that is real. It is our intuitive knowing. The energies of this transition are subtle still, but the trend that is emerging is becoming unmistakable for those who are sensitive. The goodness, the regeneration and the richness of life that we are entering energetically is so profound, that it is changing every aspect of our lives. The Earth is being renewed, and human abuse of nature is ending. All is being purified and returning to its divine design.

No longer are we being forced to work as slaves to negativity in jobs we do not want or to invest in projects that negatively impact others and other life forms. Our hearts are prompting us to pay attention to our inner guidance and to recognize the abundance that is becoming available for us to participate in the renewal of life. We are incredibly expansive Beings, having unlimited creative powers that we can realize, once we are able to trust ourselves implicitly. The timeline of focusing on our ego-consciousness to guide us through life is not our only choice. We have the option of expanded consciousness by opening our awareness to the higher energies in our conscious essence coming through our higher chakras.

While experiencing everything that is happening around us, we can practice inner awareness. Our heart has only acceptance, compassion and support through the conscious life force that constantly enlivens us. If we can align with this energy, even in the face of outer drama and threats to our well-being, we are protected and empowered in Light and Love. This is not casual energy, and the desires of our heart are not frivolous. Our essential energy is powerful and life-sustaining and enhancing. When we initiate the process of transcendence, we can stand tall and be confident in the unconditional love that flows through us from universal consciousness and enables us to live miraculous lives in gratitude and joy. This is our destiny and is the quality inherent in the new era of human history.

The Transformation of Humanity

From the quantum sciences, as well as from higher mathematics and esoteric spirituality, we can learn that consciousness is everything. There is nothing apart from consciousness, and consciousness includes awareness and realization. We have also learned that everything that exists is a fractal of something greater, and all fractals contain minute fractals of themselves, which contain even more minute fractals of themselves, and all are conscious entities, having awareness and realization. Everything that exists is conscious, from sub--atomic entities to entire universes. All are included in universal consciousness, which we may know as the essence of the Being that we may call the Creator.

As fractals of the Creator, we have all the attributes and abilities of Creator consciousness. In our deepest awareness, we contain everything from viruses and protozoa to stars and galaxies and beyond, but in our incarnation as humans, we have no grasp of our own vastness, and this is intentional on our part. We chose to experience our limitations. In this lifetime, many of us chose to be here in order to transform the consciousness of humanity, because humanity has been on a destructive course and is in dire need of a course correction.

It is through our ability to modulate the energies in our awareness that we can change the consciousness of humanity. What we do physically can be helpful, but it is what we do in our consciousness that is what shifts the polarity of humanity to life-enhancement. We must be the ones who can realize living in gratitude and joy for no other reason than that we know this is how we truly want to be. This kind of envisioning and emotional expression purges us of our limitations and creates clarity in our perspective. Everything we do can arise from the energetic level of gratitude, compassion, love and joy. In this perspective, we can exercise and care for our bodies, we can dance, sing, go shopping, interact with one another and enjoy being in nature.

We can establish a way of being mentally and emotionally that is life-enhancing in every way. By aligning with our inner knowing of what we love the most, we can transform our lives beyond the wildest dreams of our ego-consciousness. We are here to transform ego-consciousness into limitless awareness and unconditional love for the enhancement of all life, beginning with ourselves. Only by actually living in gratitude and joy can we understand how higher consciousness works. It elevates everything in our experience. We can no longer work in meaningless or destructive jobs or live in poverty and oppression. We can attract as much abundance and love into our experience as we can allow. Our consciousness has no limits, and when we can thoroughly accept this in our realization, we transform our lives and the lives of everyone around us.

Making a Leap in Consciousness

In learning to be aware of our inner guidance, we can choose to transcend ego-consciousness by eliciting the higher-dimensional energies of gratitude, joy and compassion. The vibratory spectrum of these feelings and their way of being is the kind of energy that comes through the conscious life force that flows through our heart and our higher chakras. Our guidance is what excites us most at this vibratory level. Our guidance is the energy we are being attracted to as much as we will allow. If we do not resist it through our limiting beliefs about ourselves, it draws experiences into our lives in alignment with our gratitude. Without our resistance, fulfillment is our natural state of being.

When we can direct ourselves into a state of gratitude and joy, opportunities arise for our participation with mutual love and delight in our encounters. Everything we need comes to us when we can realize its vibratory quality in our attention. We are naturally fed well, clothed as we desire, housed beautifully and transported in style. This can happen when we use our consciousness for the enhancement of all life. Taking the initiative to maintain an emotional and mental state of gratitude, joy, compassion, acceptance and love gives us access to greater realization and opens up our experience of magical living.

When we have confidence in our creative ability, we receive flashes of insight and knowing. Things happen in alignment with our own state of being. The qualities of energy that we pay attention to and align with are the qualities that manifest in our experiences. Only our limiting beliefs can keep us from realizing fulfillment for ourselves and for everyone. Our lack of Self-Realization can keep us from being our true Selves. This is our great challenge. In transcending our limiting beliefs about ourselves, we must change our attention from stress and lack of fulfillment to the enjoyment of what we love.

Making this leap in conscious realization happens in conjunction with our ability to follow our most attractive guidance. We can be aware of what feels most fulfilling in every moment, guiding us in every way. We can train our subconscious self thoroughly by much practice in learning to realize our infinite presence of awareness and creative ability. We are inherently masters of this dimension and any other that we want to participate in. Our Creator has endowed us with everything we could need and want in order to participate in experiences of all kinds in the expansion of universal consciousness. We get to choose how we want to experience our lives as expressions of Creator consciousness. We are the Self-aware ones, the ones with infinite creative ability.

Living in Our Increasingly Intense Energetic Environment

Because we have had deep resistance within our ego-conscious mind to allowing our innate divine life force to guide us, we have struggled with ourselves in our endeavors to manifest our desired creations. We haven't been able to allow our natural creative power to work properly. Living with the ego-conscious part of ourselves as our master, we've become comfortable in the world of duality. This does not allow us to realize anything greater. From within this portion of our consciousness, we have acquiesced to these limitations and have incorporated them deeply into our memory.

Now the increasing intensity of our energetic environment is causing our own ability to stay in limited conscious awareness to destabilize. For humanity, life is becoming increasingly chaotic and difficult, with the threat of world-wide break-down of our civilization. It is as if we are living energetic capacitors, with our positive charge building in intensity, until it overwhelms our ego-consciousness with realization of our infinite presence of awareness. This realization opens us to make a leap in consciousness that transforms our lives.

As we traverse our inner journey of Self-discovery, we become aware of the quality of our own mental and emotional state of being. We are beyond time/space and fully aware of our human presence. We also are aware of the etheric presence of other conscious beings. We feel their energetic signature, alerting us to the quality of their thoughts and emotions.

In our self-awareness, we have the freedom to feel however we want. We do this with our attention and mental and emotional alignment in energetic resonance. How we feel in the moment has been directed by alignment with our attention to the polarity and frequency of our encounters and environment. We can have this awareness while being in alignment with anything we choose. We can freely choose our own state of being, according to the positive, high-vibratory expressions of our intuitive knowing.

When we make the connection to our inner knowing, we can practice and gain confidence in opening ourselves to it fully. This process allows us to understand that our limitations are part of our self-created ego-consciousness. We can now expand this spectrum of energetics into greater consciousness, until we realize that we are aware of our creative essence, and we can know we are the creators of the quality of every experience in our lives by our modulation of the energies of our own vibrations. Our thoughts and feelings about ourselves express themselves as energetic vibrations that radiate from our conscious presence into the quantum field of all potentialities for manifestation. By our state of being we attract energetic patterns of experiences in alignment with us.

When we realize that our presence of awareness is beyond doubt and fear, and we know how to express the energies of life-enhancement, we can realize we are free of any requirements outside of ourselves. We can follow the expressions of what gives us the most gratitude and joy. This is what we feel most attracted to and what we most love.

Creating the Life We Truly Want

We can accustom ourselves to living intentionally in good vibrations. As soon as we open ourselves to it, we already know our intuitive guidance. We can have full awareness of the energy constantly flowing to us in our conscious life force. It can feel like life-enhancing love, compassion and joy. It inspires musicians, artists, poets, mathematicians and inventors. When we are clear and beyond self-limitations, our inner guidance is the feeling that we are most attracted to. Until we achieve this state of Being, we can use our imagination to align with our most elevating feelings and visions.

From the energetics of duality to a dimension beyond polarity, we are living in the time of change, and we are being attracted to transcend our ego-consciousness and our personally-limiting beliefs. We are being guided to live in love and confidence, beyond fear and doubt. Our social, family and cultural expectations can keep us from expanding our awareness, but we have absolute control over the focus of our attention and energetic alignment. It's as if we're playing a game of strategy in finding clues all around us and following them as we awaken to what we truly know within. Our guidance is what remains in our awareness after we have eliminated all of our personal dramas and limiting beliefs about ourselves.

Until we can resolve and transcend the extraneous attachments to our ego-consciousness, we restrict our intuitive awareness. Even though we are designed to be natural creators, we have doubted this ability and disabled ourselves. We have imagined that we are subject to conditions and circumstances outside of ourselves, and that we need to receive sustenance from others. This is all imaginary. We create our reality just by our being present in any moment. The vibrations of our imagination and emotions attract compatible, resonant energetics and circumstances for our experience. The circumstances of our lives come about as a result of our beliefs, thoughts and feelings about ourselves.

There is nothing esoteric about this. These are simple energetic principles of interaction. Our energetic modulating ability is always operating and radiating the expressions of our psyche into the quantum field, forming our experiences. When we are completely clear and present in our attention, we are in a neutral space. When we have thoughts and feelings, we are creating patterns of energy that are positive or negative and somewhere in the range from deep depression to ecstatic excitement. Where we are within this range of energetics is our personal choice, regardless of what we may be encountering in the moment. We can learn to be creative in ourselves for no reason other than to experience what we love the most in every moment. If we make a habit of paying attention to the most elevating and life-enhancing thoughts and emotions that we can elicit within ourselves, we can keep expanding into better and better conditions, until we realize that we are in a higher dimension of living.

Realizing Our Eternal Self

From within our ego-consciousness we are being attracted to the energies of great vitality and abundance. Our natural state of health is living in our prime. When we no longer believe that we must grow old and decrepit, we can open ourselves to living as long as we want in a 25- or 30-year-old body in the shape and condition that we desire. Because we have believed that this is not possible, we have made it so for ourselves.

In the same way, we have believed that we can be controlled by others, whom we do not want to control us. It is our inadvertent submission that allows this. As long as we harbor fear, we are negatively polarized and cannot imagine the reality of personally living now in a world where we are naturally provided with everything we could want, and where we encounter others who are intentionally loving, compassionate and joyful. Yet this is our natural condition, and can be our reality when we realize that it is. It is a leap of faith to open ourselves and fully want to participate in life beyond the confinements of the belief system of humanity.

We can be guided by the most life-enhancing and enjoyable feelings coming into our realization, along with inner visionary and audible impressions. Our natural condition is attracting us to our true selves, who are expressing the energies of living in unconditional love and great joy in gratitude for our Self-Realization of our eternal presence of awareness and infinitely powerful creativity. In resonance with our intuition, our mental and emotional abilities are beyond measure. By being in the vibratory range of joy, we are aligning with our true Being, our soul-consciousness. In this realization we are unlimited in every way.

When we can be mentally and emotionally clear, we are sovereign Beings, free to be however we feel guided to be. This guidance comes through everything that we identify with. By aligning with its essence, we can be aware of it. When we align with its vibrations, our intuitive guidance is what we deeply know. As we learn to practice being in positive, high vibrations in the moment, we gain the ability to transcend ego-consciousness. Our limiting beliefs become unbelievable in realization of our eternal presence of awareness.

How powerful we can be with our creative ability depends upon our clarity and intuitive alignment. Our current condition is irrelevant, because it is a manifestation of our past beliefs and expectations. As we open ourselves to a new way of being, the old must fall away. When we are in alignment with our intuition, we realize that everything we need is provided for us, and we can give and receive freely by our inner knowing. When we deviate from our inner knowing, we are subject to the limitations of ego-consciousness. When this happens, we can intentionally re-orient ourselves to Self-Realization of our eternal presence of awareness and infinitely creative consciousness. We are designed to be the masters of our dimension in unconditional love and joy.

Transforming Chaos and Hatred into Compassion and Joy

Accepting the chaos and destructive energies in our world without resistance is a necessary step in our own transformation and the transformation of all of humanity. Our natural state of being is thankful, open, joyful and confident. It is with an elevated state of Being that we can accept the negative energies that are manifesting among humanity. While giving them only passing awareness, we do not need to align with them in resistance. We can accept them in neutrality, and balance them through our alignment with the light within, allowing for acceptance in neutrality and transformation into energies that are life-enhancing. We can do this in our own awareness and can relate with the higher aspect of any pattern of energy or any person.

The life-force is always present in everything that exists and is part of universal consciousness. If something is so evil that it has no light within, it ceases to exist. Negativity is ultimately self-destructive. It needs our life force through our engagement and alignment with fear. Without our resistance to evil, we give it zero energy. Our focus on the light within every encounter enables it to appear for us, and we can realize the higher energetics involved everywhere according to the vibrations that are present.

We have the free choice to focus on whatever we desire to interface with. We can allow our realizations to come through our imagination and our emotions and then to fill our awareness with their reality. This is how we can free ourselves from our limitations and transform ourselves and humanity. When we are living in gratitude and joy, negativity cannot exist in our experience. What is happening around us is irrelevant. Those conditions change with the consciousness of humanity, just as does our condition with our attention and alignment with the emotions of our heart.

By changing our own perspective and way of being, we change the quality of our experiences and what's happening around us. When we choose a higher vibratory life, this is what we are creating for ourselves. We can transcend negativity by paying attention to what gives us gratitude, compassion and joy, while being open to great wisdom within. We can accept all the characters in the drama around us with neutrality. In our own state of being we can transform negativity into neutrality by our own awareness of the balancing energies of life-enhancement. Without our alignment and engagement, negativity is neutralized and disappears from our experience. We can continue to pay attention to the vibrations of our heart in gratitude, compassion and joy for all of our experiences.

Following Our Inner Light

On the inward journey, we may face many difficult obstacles in freeing ourselves from fear of transcending ego consciousness beyond the body. When we want to know our true Self and realize who we are, our search can be within ourselves to what we truly know without knowing how we know. We can realize our own essence beyond time/space. We can be aware of ourselves as our presence of awareness with creative mental and emotional abilities. We can authoritatively control these abilities in alignment with the feelings that come from our heart. As we live in awareness of these vibrations, we can be confident in this kind of knowing beyond ego-consciousness and expect that our outer lives will come into resonant alignment with our inner knowing.

Everything that happens in our lives results from our mental and emotional creations. We express vibrationally what we feel and think about. This is how we're designed. The vibrations that we express are our personal energetic signature. The resonance of our energies attracts situations and people that are compatible with our mental and emotional states.

We can learn to keep ourselves in alignment with the positive, high vibrations of our intuition and inner knowing. By changing our attention from thoughts and feelings tinged with fear to their transformation in joy, we change our condition and experiences.

What holds us in the realm of duality is fear in all the forms of its energetic spectrum. Fear can be neutralized by realizing our essence beyond the body. The threat of leaving the body is meaningless. We leave our body-consciousness all the time in day-dreaming, night-time dreaming and perhaps being aware within of not being anywhere. Being in the body is a virtual experience that we have perfected as reality. In its essence the physical world consists of electromagnetic wave-bands of frequencies that we align with in our consciousness, interpreting the polarity and vibrations as stimulation of our senses and seeming to be solids, liquids and gasses.

By reacting to negative encounters with compassionate understanding and intentional awareness of the light within everyone and everything, we transform our situations into experiences of gratitude and joy. This is our natural state of Being, when we have resolved and transcended our limiting beliefs about ourselves. Our heart guides us to think and feel unconditional love as our natural state of Being, along with beautiful visions of how we want to live. By imagining living in the consciousness that expresses these energies, we create their reality for ourselves by realizing what they are.

Creating Our New Reality

We experience what we believe is real. If we change our beliefs about reality, our reality changes to align with our beliefs and realizations. The empirical world begins by arising from consciousness and takes form by attracting and organizing all of the entities that make it a physical experience in a range of resonating vibrations that we can feel and perceive. It is a manifestation of consciousness, ours and everyone else's, who is participating in this dimension. With our imagination and realization we provide the modulation of the energies that we focus upon and align with.

In our personal lives we direct the qualities of our experiences with our state of being. Our recognition and belief in the qualities that are present in our awareness in any given moment are the creative impulses that bring us our experiences. The only difference between something that is real for us and the same thing that is imaginary is our belief and realization. The energies themselves are the same, and they attract experiences in the same way.

We can use our imagination in alignment with the energy of our heart. By being in gratitude and joy, we can imagine every encounter, whether with nature or with people, as an exchange of love and light. This will create its reality in our experience. When we can be this way, we transform our lives and live in a dimension beyond duality. When we are heart-centered, we love to be in each other's presence.

If we have negative experiences, we can resolve them by continuing to interact with whatever level of light may be present in everyone involved. Making our way through the experience with intuitive guidance, we can stay in gratitude and compassion. All of this is a leap beyond ego-consciousness, which we can resolve within ourselves by paying attention to our inner knowing and aligning with the feeling of our heart-consciousness.

There are many ways to mastery of human life as we know it. Many spiritual masters in the past have shown us the ways. Their message was always the same, that we are masters who have been unaware of who we are. Our intuitive knowing opens our awareness to a greater life that is a conscious expression of unconditional love, joy, abundance and freedom in a dimension beyond negativity.

Increasing Our Inner Radiance

In our quest to know the truth of who we are, we can raise our vibrations by choosing to fill our awareness with greater gratitude, appreciation and joy. The further we can go within ourselves into selfless and self-fulfilling love, the more present in multi-dimensional awareness we can become. Our condition flows energetically with the vibrations of our imagination and emotions. To be fully open to realizing our limitless awareness, we must drop all of our preconceptions and limiting beliefs about ourselves. We are pure consciousness with present awareness wherever we direct our attention, and we can direct it to ourselves. We can realize our essence as fractals of the Creator of all, with inter-dimensional awareness and creativity. We set our own limits in all aspects of life.

Without psychological limitations, we are free to follow the vibrations of our heart, leading to intuitive knowing and joyful confidence. When we are able to maintain this awareness, our lives become magical, with everything going smoothly. If we can maintain this awareness even when faced with suffering and pain, we transform negativity into well-being and love. If we let ourselves be distracted by even a small amount of fear or doubt, we cannot fully open ourselves to heart-consciousness. By following our heart-consciousness, we can have absolute confidence in everything. We can clearly know our guidance in every moment.

Our experiential world is an unbounded sea of electromagnetic wave patterns, all of them arising out of universal consciousness, which we participate in, as much as we allow ourselves. When we change our consciousness, we change the vibrations of the wave patterns that we pay attention to. This is how we create the qualities of our experiences. Our lives are completely an inside job. We direct the energy in our consciousness with our attention and choice of polarity and vibratory level. This we can learn to do intentionally all the time.

When we have resolved our personal dramas, needs and desires that are all based in fear, we become able to pay attention to our heart-consciousness. It has the vibratory level of gratitude, compassion and joy. When we can hold our awareness to this spectrum of energetics, we transform our lives and can realize our true creative power. We can realize the reality of this range of vibrations, and, by constant awareness of the energy of our heart, shift into a dimension that cannot be reached by negativity. We can fill our awareness with love, joy, abundance and freedom as we expand into infinite presence of awareness.

Opening Ourselves to Greater Awareness

The energy of our heart-consciousness vibrates within a dimension beyond the awareness of our ego-consciousness. This is the reason our ego-mind cannot allow us to realize our true essence. The ego is an artificial entity that we have created to express aspects of ourselves that we choose to experience. In life-enhancing awareness, we can resonate with gratitude and joy. In this dimension, we can be our human presence, while living in eternal presence of awareness beyond limitations. This realm is invisible to ego-consciousness, but we can satisfy all of our ego's fears and needs by opening ourselves to a higher vision of life that we can invite into our awareness. We can realize more light and love in our experiences.

We can find what we love the most by realizing that it is already ours, because it is. Whenever we recognize and realize that it is present in us, the energy we love comes into our experience. We can experience its energetics in every encounter of any kind. Our attentive awareness manifests its reality for us. What we love the most is the energy of our heart-consciousness. Being in resonance with its vibrations can open our awareness to a more expanded level, a higher dimension. The desires of our heart-consciousness are the thoughts and feelings that are life-enhancing in every situation. Our intuition can guide us there.

Like our awareness, our creative ability has no natural limitations. Any limitations are self-imposed and can be resolved. First we have to realize what our limiting beliefs about ourselves are. They are all negative and are based in fear and doubt. We can transform these energetics by shifting our attention to the expressions that we love, opening ourselves to an expanded awareness. In this way we can align with the light in any situation.

If we can achieve constant positivity with acute awareness of our intuitive knowing, we can live completely fulfilling lives while living as ordinary humans. In our experience we can understand everything that is occurring in our presence. This enables us to transform our energetic environment into a radiance of life-enhancing visions and emotions. When we finally realize that we are everything we could ever want, we free ourselves to align with our true essence in infinite Being.

In alignment with eternal Self-Realization, we can open our awareness to a realm of brightness and joy. Fulfillment of the desires of our heart-consciousness naturally comes to us when we recognize and realize the presence of love in our experience. At this level of attention, we are in the realm of life-enhancement.

We have the choice of paying attention to anything that we can be aware of. We choose our energetic environment with the vibratory level of our thoughts and emotions. Our attention is uniquely ours and is the essence of our creative ability. By aligning with the intuitive knowing of our heart-consciousness, we can live in gratitude, confidence, compassion and love. While maintaining these qualities of being, we can open ourselves to our eternal presence of awareness with infinite creative power.

Aligning with Our Inner Self

Each of us constantly receives a flow of conscious life force that has the energy of life-enhancement. We are free to use it as we please. It works by vibration and polarity. We receive our life force through the heart of our Being, and we can invite its feeling into our awareness. It is the source of our inspiration and everything that we deeply know, and it flows with unconditional love. When we can identify its vibratory quality with our attention and emotions, we can intentionally align with it. Living at this vibratory level transforms our awareness into a dimension beyond duality.

With an intentional desire to be grateful, appreciative, compassionate, loving and joyful, we can open our awareness to a realm of beauty and fulfillment, regardless of any drama of negativity around us. From its spectrum of polarity and vibrations, negativity cannot interface with us in a higher dimension. Only if we engage with negativity can it influence us. It needs our life force, which we provide with our attention. We pay with our attention.

By paying attention to our own inner energy, we can come to know ourselves acutely. In loving ways we can connect intimately with our subconscious, innate being, which connects us with universal consciousness. Its messages to us are based on innate knowing of everything. They do not contain or result from inductive reasoning. For our ego-mind this is all a mystery.

Because ego-consciousness is based in fear of termination, it has no existence in our eternal presence of awareness.

We can open our awareness beyond our ego-mind's limitations by paying attention to the qualities of our thoughts and feelings. In every moment we can choose to align with life-enhancing vibrations. This vibratory level is beyond fear and negativity. They are present in a lower dimension. In the realm of joy and love, we are transformed in alignment with high vibrations. Positivity fills our awareness and brings us fulfillment of the desires of our heart and everything that we imagine and love.

Because we have limited ourselves in every way, except for our intuitive knowing, we need a strong intention to transcend the limitation of ego-consciousness. To the ego, intuitive guidance may be unbelievable in many ways. We are being challenged to be sensitive to what we deeply know and to trust that it is true. Once we can realize what our intuition is, life becomes easy, and we can feel fulfilled in being able to create every experience we could ever want.

The Magic that Yields Ultimate Freedom

We are all capable of injecting magic into our lives. This is our opportunity to stop being lightweights, compared to our real abilities. We no longer have to play sick or poor or enslaved on whatever level. Our conditions are entirely our own vibratory creation. We are our own prison guards. Regardless of how we arrived in our present life experience, our lives from now on can be completely different. Our experiences are created in universal consciousness by our own thought-patterns and emotional state of being. We can shift our focus of attention to the vibratory spectrum of our heart-consciousness. The heart leaps to our support whenever we need more oxygen, regardless of what we use it for. This is unconditional love in personal freedom and is the level of vibration that we can resonate with in every moment.

When we allow ourselves to align with the energies of our heart-consciousness, we enter a realm of goodness and abundance without limit. Our heart-consciousness supports our bodies in the best ways that our general consciousness allows, as expressed in our energy signature. When we open our awareness to life-enhancement in every way, we can feel what unconditional love actually is. We are in a realm beyond ego-consciousness. Now we're beginning to gain some stature as real humans.

The quantum field of all potentialities holds every possible kind of experience. Our consciousness sources our personal experiences from the vibrations of our thoughts and emotions, expressed as our energy signature. Every experience is a result of the operations of consciousness, including our attitude and perspective. We are constantly creating a flow of conscious and subconscious vibrations that we radiate from our essence. The vibrations that we choose to hold in our imagination and feelings establish the frequency band and polarity in our psychic environment, which we realize as our reality.

While practicing living in the energy of our heart, we can enhance life all around and within us. We can align with experiences that have this quality. This is when magic occurs. It is what blows ego-consciousness away. Living in the vibrations of joy, compassion and life-enhancement, we transcend the limits of ego-consciousness. Our situation comes into alignment with our elevating and expanding energy signature. Our circumstances arrange themselves to manifest the energetics that we create in our alignment with heart-consciousness. We enter a dimension of mastery with our mental and emotional abilities in service to unconditional love in the consciousness of the Creator. We can come to realize in our essence that we are the Creator.

An Enlightened Life

What can we do with ourselves, once all our needs are fulfilled? Our inner guidance is steering us toward life enhancing expressions in every way. By enjoying ourselves in pursuit of our passions, we may have desires to be inspired in new ways. We can maintain a perspective of gratitude, compassion and joy. In our endeavors we may feel desires from within to create something or take some kind of action or just to be present in a high state of vibration. This is where inventiveness and creative visions arise. There is much that we can do to brighten the life of humanity.

All are our brothers and sisters, and all arise within the consciousness of the Creator as fractals of the entirety of universal consciousness. Although we can describe it, our ego-consciousness can grasp only part of our enormity. We have the ability to give without limit in every possible way, enriching our own experiences in the process. Our sustenance is our consciousness and our ability to create a high-quality of life effortlessly by intentionally living in positive, high vibrations in alignment with the energy of our heart-consciousness. In this state of Being, we always know everything we could want to know about our situation and anything that we pay attention to. When used in alignment with our heart-consciousness, our free will is our eternal gift to use our awareness however we desire.

When we are in vibratory alignment with our heart-consciousness, we can direct infinite energy through our conscious expressions to elevate the hearts of humanity. We can become powerfully radiant with unconditional love, compassion and joy. Our Source of every expression is the consciousness of the Creator. We are conscious extensions of the Creator. When we live in the dimension of unconditional love, we enjoy infinite creative power, while choosing to respect the sovereignty of all other conscious beings.

Without interfering with the consciousness of anyone, we can help to provide an environment of abundance and freedom for all. Once we're flowing with this energy, we always have more that we can create. The more creative we become, the more fulfilling our lives become. We can become the masters of our attention and emotions, our visions and desires, with confidence in our creative ability without end. We can make our lives as much fun as we want, with as much excitement or solitude as we may choose.

When we enjoy ourselves with love for all, we expand the conscious awareness of humanity. We can be especially helpful in showing the poor and depressed that they can free themselves from the prison they have held themselves in. There are no needy people on Earth. There are, however, many who have believed themselves into servitude. That includes the vast majority of humanity. When they realize that we are truly free, they want to know how to do it. This is the opening for the Light to enter more hearts and for greater expansion of heart-consciousness.

Our Identity beyond Ego-Consciousness

When we are in beautiful and spectacular places in nature, we feel a natural gratitude for the beauty and magnificence surrounding us. We feel a heart-consciousness connection. Inspiring music also elicits this energy within us. Gratitude enables us to transcend our ego-consciousness. Once we make the leap beyond duality and into our heart-consciousness, we have left our ego-consciousness in our memory, along with all negativity. When we pay attention to our heart-consciousness, we give no conscious life-force to negativity, and it dissolves from our awareness.

In a higher dimension, negativity cannot exist. Its vibrations become unstable, and negative consciousness loses its ability to express itself. It dissolves into a lower dimension. In its place we can experience the vibrations of our heart-consciousness in gratitude, compassion, joy and appreciation. This vibratory level connects us with universal consciousness, filled with infinite love and life-enhancing expressions of beauty and majesty. We can expand with this energy as far as we are willing to go. The more we expand, the more intense everything becomes, and our creative power strengthens greatly.

No longer are we subject to any controlling powers in the spectrum of human experience. We are sovereign Beings of eternal present awareness, able to express ourselves in any form or entity we desire. By living in the energetic band of frequencies of unconditional love and life-enhancement, we are beyond the limitations of negativity. This is the realm of ascension into a higher dimension of living. This dimension exists right where we are now. It's just a different, positive vibratory rate. In our awareness, we can inhabit both dimensions and more, depending on our personal degree of clarity.

The difference in the quality of our personal experiences between the higher and lower dimensions is extraordinary. The lower dimension contains all of the expressions of negativity. The higher dimension is fulfilling in every way, leaving no personal drama or needs. All we need to do is to be ourselves on a heart-conscious level, and every desire of our heart naturally receives fulfillment. We can have complete confidence in our lives and realize our infinite creative ability, benefitting ourselves and life all around us. We become radiant expressions for the light of our heart.

Resonating with Heart-Consciousness

In our own awareness we can choose to receive everyone in the most loving and joyful ways. In our perspective we can encourage everyone to pursue their own interests and passions in heartfelt ways. Ego-centric interests are self-limiting and are now being dissolved, along with the entire realm of duality. It is all fading into its own dimension, apart from heart-consciousness.

Awareness of heart-consciousness can result from subconscious prompts at critical times, opening our awareness to greater insight and wisdom. We have always had this awareness, but our ego-consciousness has hidden it from us by doubting it could be real. Doubt must be resolved through recognizing its energy patterns, its polarity and emotional basis. When we can feel the nature of doubt, we can understand its basis. It is based on an imaginary state of affairs that we have created with our thoughts and feelings. It has no reality of its own. It is strictly an expression of our imagination, based on the fear that we can be inadequate in ourselves. It is based on the belief that we may not be eternal in our personal awareness.

This belief is purely a personal choice, which we can change at any moment. Is the requirement that we live in doubt and fear anything but hearsay? How can we know what we do not know? Knowing requires an intentional choice to open our awareness to our heart-consciousness. We can ask our subconscious to help us in transcending ego-consciousness by aligning with energies that we love and are grateful for.

In the radiance of our eyes, we express our heart-consciousness, and we feel it as inner joy and warmth. When we choose to live in constant awareness of our heart-consciousness, beyond the limitations of the ego, we can realize that we are in a dimension of love, compassion and joy, which we share with everyone else who is in the same vibratory resonance. We can recognize each other in our eyes, and we can feel each other in our heart. This is a dimension that we enter and experience fully, when we realize its reality. It requires us to open our awareness to it and align with its vibrations. We can search universal consciousness for the vibrations that we want just by desiring them, and we will be aware of them.

Our awareness provides the conscious life force that results in the expression of our energetic patterns into our experience. When we are filled with joy, we create joyful experiences for ourselves and everyone in our energetic aura, and we expand the consciousness of humanity, along with our own. When we can be clear in our awareness and transcendent in our presence, we transform ourselves and the energetic expression of all of humanity. We all have the same essence of consciousness, enlivened and filled with the light of the Creator and destined to attain clear Self-Realization.

A Path to Personal Transformation

From quantum physics we know that there is a universal consciousness that all conscious beings participate in. We all have infinite consciousness as part of our essential Being. Although our awareness of it has been removed, we can regain it. It depends upon our personal interests, our understanding and our intentions. Our greatest tool in expanding our consciousness is our free will. We have absolute freedom in choosing what we pay attention to. What kind of energies fill our awareness is our personal choice. In the dense energies of the empirical world, we have powerful distractions, but they all are only energetic wave patterns that require our realization in order to affect us. We make them real in our experience by our belief in their reality. Our realization gives them access to our life force in resonance with them. Our senses become receptive to their energetics. It is all within the absolute direction of our awareness and realization. Being able to direct our realization intentionally in every moment requires practice.

When we have an interest in understanding life on a deeper level, we open ourselves to it and become receptive to the signs and symbols around us, leading us to realization of our own essence and abilities. When we focus intently on our own nature, we find nothing material there. All we have is our awareness of ourselves. We can have awareness without limits. This is where we must be intentional.

Realization comes as a result of our choice of using our attention in creative ways. If we turn our attention within to our heart-consciousness, we come to our eternal presence of personal awareness. Our physical body is an expression in form and substance of the perfect vehicle to experience the energetic patterns that we chose to attract, consciously and sub-consciously. This happens in the conscious awareness that we choose to give our attention to.

If we choose to realize more light and beauty in every moment, we come to expect these energies in our experiences. The current human environment can be transformed in our own realization of the energies and forms of beauty and joy within ourselves. In our experience we can fill our awareness with the emotions and visions that we feel in our heart-consciousness. This is where we can be truly creative in imagining magical situations and environments, new technology and techniques for creating virtual realities expressing heart-consciousness.

In our own awareness we can realize our mastery of the empirical world as a result of learning how to project our conscious realization. It begins with directing our attention to feelings and visions that we want to experience. We develop our personal energetic signature by our state of being, by the qualities of our thoughts and feelings. By intentionally aligning our mental and emotional processes with our heart-consciousness, we raise the frequency of our energetic signature and transform our experiences in the world.

Experiencing Levels of Realization

Realizing who we are and what our essence is can help us to live the kind of lives we were created to enjoy and participate in. Creator consciousness is constantly expanding and enhancing all conscious beings. To the extent that we have eliminated Creator consciousness from our awareness, we must live in ego-consciousness in the awareness of duality and degradation through entropy.

We can learn to read the energy of everything in any situation. We can learn to transform negative energy to its positive counterpart or resolve it into another dimension. Our situation can consist of grateful and joyous experiences in interaction with all that comes into our awareness. We can direct these feelings to our awareness, establishing a high-vibration state of being. In any situation we can intentionally interact in alignment with our heart-consciousness, knowing that we have infinite creative power.

Threats to our well-being or so-called accidents do not exist in the heart-consciousness dimension. We can be completely and intentionally creative in all aspects of life. We can learn to direct our attention entirely to the kinds of situations that we love. As we pay attention to how we want to imagine everything that we are aware of, we can intentionally modulate the energetic patterns around us into alignment with our intuitive knowing. We can be affected only by energies that we accept into our awareness and pay attention to. We are energy modulators and experience creators. This is what we do all the time, whether we are aware of it or not. We are constantly creating the qualities of our experiences by how we choose to be in every moment.

Where we are and what we are doing matters on an energetic level, but not necessarily on a physical level. It is our vibratory level that determines how smoothly our life goes. It is not important if we have a job or a business or any socially-acceptable style of life. What matters is the vibratory spectrum of our energetic signature. Our sustenance comes to us naturally at the level of our vibrations. This level is completely subject to our personal direction and choice. How we choose to feel in any moment is important for our creative direction.

When we are Self-Realized, we are infinitely powerful creators with our intentional visionary and emotional abilities. We can realize that the experiences that we love the most are the ones in alignment with our heart-consciousness. With all of our needs fulfilled, we are free to create a world of experiences that express the energy of our heart. Since these vibrations are unknown to ego-consciousness, we must open our awareness to heart-felt feelings and thoughts about everything and everyone. Heart-felt energy is life-enhancing in every way. It is true creative energy that we can be aware of within our greater consciousness.

Working with Our Energetic Alignments

On the lower astral level, this planet has been controlled by an empire that encompasses nearly all of humanity. It is an empire based on control through fear and monetary management. It controls all of the militaries of the world, all of the underworld mafia, weapons dealers, human trafficking, banking and monetary systems, all multi-national corporations as well as all governments and politicians. What is the purpose of the empire? Why does it need to control everyone? It is an agent of darkness and an expression of duality. It converts the life force of humans into aspects of commerce through trickery and dissimulation. These commercial activities provide the basis of life for the psychopathic rulers of the empire, who have terminated their connection to their natural flow of conscious life force. The empire is parasitic in order to express deeply negative energetics that have no origin or support from Creator consciousness. The empire extracts its conscious life force from humanity in order to express self-destructive energetics that diminish all life everywhere.

We have become so entranced in the program of the empire, that we have sold our essence through our attention and alignment with fear, which we have created for ourselves as directed by our controllers. Fear originates in the belief in our mortality. This is an artificial concept that exists only in our psyche through our realization of it as real. If we did not believe in our mortality, it would not exist. What is mortality, if not a belief? Our consciousness and present awareness are the basis of everything that we experience.

Unless we believe in its reality, the empire cannot affect us. If we are beyond fear, we are expressing no vibrations in the spectrum of duality. We can free ourselves with our vibrations. Any enslavement that we feel or accept is purely voluntary on our part. By complaining about or fighting against negativity, we give it our life force through our alignment with its range of polarity and vibrations. We engage with it. If we remove ourselves from duality through our attention to the life-enhancing energies of our heart-consciousness, we can use our mental and emotional abilities to enter a higher dimension that is fulfilling in every way.

Deeply set in our consciousness, our limiting beliefs can dissolve when we change our perspective and disbelieve in duality as real for us by paying attention to what we truly love and want to experience. By opening ourselves to our heart-consciousness and aligning with it in our imagination and emotions, we come into mastery of the human situation. We can more easily open ourselves to our higher chakra energy centers, realizing our greater abilities of telepathy and more.

Identifying Our Essential Self

We are so awesome, so far beyond ego-consciousness, that we cannot even imagine who we are in our essence. We can rise to the level of spiritual mastery, materializing things we want, healing the sick and wounded and even having fun, such as when Sathya Sai Baba would have the stars twinkle behind him. More than any of this, however, is our ability to enter awareness within universal consciousness and identify with the consciousness of the Creator in unconditional love and enhancement of all life. In this state of Being, we are the presence of infinite awareness through our heart-consciousness and higher energy centers. This level of personal realization opens us to such amazing, heart-felt energies, that we become radiant with joy and compassionate wisdom. We can realize our absolute sovereignty and freedom in every way.

In our conscious awareness we can fill our attention with gratitude for the perfection of our life and fulfillment of all of our heart's desires. We can be beyond the confinements of any beliefs in personal limitation. This is all possible for anyone with a strong intention and willingness to practice intuitive awareness. We have many things to help us in our development, things like deep breathing techniques, soul-full singing and dancing, listening to inspiring music, and aligning with the energies of the Spirit of the Earth and all of the conscious beings in nature.

With some exceptions, we were born without conscious awareness of our true essence, and we have allowed ourselves to be programmed to believe that we are mortal and subject to the control of forces beyond our own capabilities. If we are willing to be spiritual adventurers, we can question this situation and turn our attention to ourselves to seek our true essence. Our personal truth is available for each of us to realize. It lives in the conscious life force that we constantly receive through the heart of our Being. We can open ourselves through gratitude to feel this energy. It has the quality of creation and enhancement of all life. We recognize it as love and joy.

In realizing our heart-consciousness, we can transcend ego-consciousness by filling our attention with our essential eternal presence of awareness of who we are. We can realize that we arise within the universal consciousness of the Being who constantly gives us our life essence with the freedom to use our life force however we choose. Once we can resolve and transcend limiting beliefs about ourselves, we can learn the truth about ourselves and our abilities in infinite love and creative power.

Living Beyond Fear

When our consciousness expands beyond time/space and duality into a realm beyond polarity, everything is present wherever we direct our attention. Unknowns and mysteries can be resolved through multi-dimensional quantum energetics, which we can know intuitively. The goal is to awaken our sensitivity and receptivity to the energy of our heart-consciousness. We can know this consciousness by its vibratory qualities, which are always life-enhancing for everyone.

When one person receives something, its energy patterns become available to all, because all participate in universal consciousness. We all have the freedom to recognize what something is and to realize if it is real for us. We can choose to align with its vibrations. We can choose the quality of our reality by our perspective, our imagination and emotions. Through these abilities, we modulate and form the energetic patterns of our energy signature. This is our personal radiance and is the quality that the quantum field manifests in our experience.

Our reality constantly shifts according to the forms and sensory stimulation we feel and imagine. We have imaginary visions and feelings that arise from energetic patterns that we are receptive to in the quantum field of all potentialities. As we shift our openness to positive, high-vibratory thoughts and emotions, such as compassion, beauty and joy, we align with our heart-consciousness. By choosing to be in positive resonance with heart-consciousness, we are guided in clarity and confidence throughout our life.

When we awaken to infinite love and joy, we are aligning perfectly with our heart-consciousness. This is our natural state of Being. As our eternal presence of awareness in alignment with our heart-consciousness, we arise within the consciousness of the Creator of all. It is a consciousness of love, joy, ecstasy and sovereignty in the enhancement of all life.

As long as our consciousness is limited to the physical body in time/space, we align with fear of physical suffering and termination. By opening our awareness beyond time/space, we can transcend fear and replace it with confidence in the guidance of our heart-consciousness and higher etheric energy centers. We become unavailable to negative vibrations by paying attention only to scenarios that we love and can attract into alignment with us. Through our enlightened gratitude and love, we can live in the vibrations that enhance all life and bring fulfillment in every way.

The Evolution of Human Consciousness

In the realm of duality, there must be balance between positivity and negativity. This is possible only if humanity is balanced. If humanity becomes too bright, the negative disappears into another dimension. If humanity becomes too tyrannical and enslaved, its life force diminishes until it cannot express any creative energy, and dissolution occurs into a lower life form with less creative ability. For eons humanity has chosen to emphasize negativity and destructive energy. Through lies, trickery and every kind of mental and emotional interference and management, we have been trained to believe in mortality and fear. This state of being provides no understanding of unconditional love, and it is self-diminishing through entropy as well as our polarity and vibrations.

Now, however, the energy is shifting toward positivity and brightness. Mortality, as in termination of consciousness and physicality, becomes unbelievable for all who have had out-of-body experiences. This group of humans has established the knowing of our eternal presence of awareness beyond the body and the empirical world. This knowing is already in the consciousness of humanity. Now the rest of us can intend to open ourselves to our intuitive knowing of the essence of who we are. We only need to recognize it and realize what it is. It comes to us through our alignment with the energetics of our heart-consciousness in the enhancement of all life. It is the vibratory spectrum of unconditional love and infinite creative power. This energy cannot be described in human language, but words can give us a hint of what it is. It is realization of our reality, the essence of who we are and what we can do.

There are a few requirements to fulfill along the way. Mental and emotional clarity are needed to understand every situation and scenario. This means that we must resolve our limiting beliefs about ourselves and our life beyond time/space. All of these processes work together in alignment with our intentions and ability to direct our attention. When we can focus on compassion, gratitude, love and joy, we are aligning with our heart-consciousness. This kind of attention from us helps us to open to our eternal presence of awareness beyond the body. Because this is our natural state of being, and it is being energized by all those who have awareness beyond the body, we can more easily recognize this energy and realize it is arising in our experience. We are evolving beyond duality and into a higher dimension of what we recognize as real.

Experiencing the Vibrations We Love

We are our infinite presence of awareness, with absolute control of our creative essence. We may be able to remember this state of Being from before this incarnation. With eternal awareness of ourselves, we can live beyond polarity in resonance with our heart-consciousness. We are attracted to life-enhancing energies, and we can resolve our ego-conscious need for stimulation and attempts to defy mortality. Once we realize that mortality is only a belief that accompanies fear and doubt, we can realize that we can change our beliefs about ourselves. This changes our experience of life. We may even learn that we can be beyond beliefs in our awareness. There are many ways of doing this. One of them is intentionally opening our awareness to our heart-consciousness. This awareness gives us confidence to be able to involve ourselves completely in experiences that we love. We can be passionately joyful and grateful for the conscious life force constantly filling our sense of Being and providing us with creative power.

Beyond the limitations of ego-consciousness, we can realize our creative ability. Without understanding our ability to modulate energy, we create all kinds of chaos and drama in our lives, because we allow our attention to be directed randomly, including to negative, life-diminishing vibrations. Whenever negativity arises in our awareness, we can realize that we need to redirect our focus to the life-enhancing energies that we love. We can recognize their symbolism in every situation, allowing us to be thankful and joyful in constant fulfillment and alignment with heart-consciousness.

With complete confidence in our inner guidance, we can live as we truly desire, among others who live with the same inner guidance and heart-consciousness. Psychopaths cannot exist in these vibrations. This leaves the parasites in a lower band-width of energy. Our relationships become heart-felt and inspiring. As we pay attention to our heart-consciousness, our personal radiance increases greatly, and we can fill our awareness with beauty and love.

As we attract other awakening brothers and sisters, we intensify one another's expanding awareness. We can recognize and feel one another's presence by the light in our eyes and the vibrations of our hearts. By our state of Being and our radiant energy, we are expanding the potential awareness of humanity. By our heart-directed thoughts and emotions, we can create a new world of unconditional love and joy for ourselves and all with whom we interact.

Realizing Our Personal Truth

Who we believe we are determines the mental and emotional limits of our awareness. Our reality exists within the limits of our awareness. To expand our awareness beyond our self-limiting beliefs requires us to pay attention to our intuition, and to do so in mental and emotional clarity with direct focus upon life-enhancing vibrations, which we can know through our heart-consciousness.

Our entire world is within our own consciousness, and we can become aware of this. We constantly shape and form energetic patterns of experiences in our imagination and emotions, creating our energetic signature of radiant energy. In resonance with our own energetic identity, we attract the experiences that we choose to recognize and realize as real within the limits of our imagination and feelings.

Our limitations are entirely self-imposed. They have no reality beyond our beliefs. Recognizing our beliefs and examining them energetically for any traces of fear or doubt enables us to realize that there is nothing there. Our beliefs are artificial constructs that we have imposed or accepted about ourselves. We have acquired them from inheritance and telepathic training. Our most powerful limiting beliefs are based on extremely negative experiences that we engaged with and suffered through.

By realizing that we allowed ourselves to align with negative energy, we can instead choose to pay attention to our heart-consciousness, transcending our limitations and opening our awareness to new realizations. Unless we give it up, we have control of the quality of our thoughts and emotions in every moment. We can command our innate consciousness, the repository of every experience we've ever had, to align with our heart-consciousness in every moment.

Beyond ourselves, there is no requirement that we ever engage with fear. Without our mental and emotional alignment, it could not exist for us. We create our own fear entirely. In heart consciousness fear does not exist. There is only infinite love and connection with the consciousness of the Creator. This has very practical implications for us, because once we realize our essential presence of awareness and unlimited ability in every way, we transform our lives and can live in a higher dimension of beauty and freedom.

By resolving our limiting beliefs about ourselves and intentionally aligning with the energetic spectrum of our heart-consciousness, we can keep creating new experiences of life-enhancement for ourselves and all around us.

Creating a Fulfilling Life in a Time of Great Challenges

To understand the nature of our lives, we may begin by learning to take absolute control of our attention. We can learn to open our awareness in every moment to our heart-consciousness and the vibrations of our higher chakras. We can learn to transcend our limiting beliefs about ourselves by resolving them and feeling their nothingness apart from our beliefs. If we so choose, we can understand ourselves as expressions of Creator consciousness through our imagination and emotions. We serve as creators of our experiences. Our vibrations are known throughout the cosmos, and we contain the vibrations of the cosmos in our inner knowing. It is the same for all conscious beings, except that we are the creators of the forms and vibratory levels of all the energetic patterns within our attention. Our attention and emotional alignment are our creative essence. Their vibratory level attracts and shapes the energetic patterns that become our experiences.

What we believe is real is made experiential for us. When we can control our imagination and emotions in any situation, we can choose our vibratory level and live in it. Our inner and outer conditions will continue as they have, until we intentionally direct our attention in alignment with our inner knowing of life-enhancement.

We are actors playing our roles in the drama of humanity.

Deep in our consciousness we know what our roles are and what we want to experience, and why. This is where our limiting beliefs about ourselves come from. They are necessary for the most convincing experiences in duality that would not be taken seriously in our infinite awareness. We know what a wide spectrum of negative vibratory patterns feels like. This experience is what we have contributed to universal consciousness, and it is what we are transforming into resonance with gratitude, unconditional love and joy.

In every moment we are free to choose how we think and feel. This decision is not dependent on any experiences we have had or may be facing. How we feel in our own being is independent of anything else, and we have control of our polarity and vibratory level. Our state of being is our creative consciousness and our personal fulfillment or lack. From moment to moment our experience is a cumulative result of our own vibrations.

By learning to direct our own level of consciousness, we gain confidence in our conscious realization and inner knowing of more than one dimension. Our consciousness is the realm of cause throughout our life. How we direct our attention and what we can realize create the vibrations that echo throughout our species consciousness. Living in resonance with our heart-consciousness brings fulfillment to us personally and also transforms human consciousness.

Understanding and Transcending Our Limitations

To know our true expansiveness, we can learn how to shape our beliefs about ourselves. We can ask if they make sense and if they are based on the consciousness of our heart or in some aspect of fear. If fear is involved, they diminish our life force and limit us. We can realize that it is possible for us to be aware of our inner knowing and to pay attention to the energies that we love and are grateful for. Our choice is always for love and expansiveness or for fear and contraction. When we are living intentionally, we can constantly choose our energetic alignment.

We are playing a psychological game with ourselves. We've placed ourselves in a psychic maze, and we're trying to find our way out. The way out is our choice of reality. We choose our reality with the vibrations that we pay attention to and align with. By imagining scenarios filled with love and joy, we open ourselves to experience those energies. By imagining scenarios filled with tyranny and adversity, we open ourselves to those experiences.

What we pay attention to is important for our personal energy signature and the limitations that we place upon ourselves. If we feel fulfilled in every way, we experience the energy that we align with. When we feel trapped in a limiting lifestyle, we stay that way until we change our belief about ourselves. When we elevate our feelings to freedom, gratitude and joy, our limitations can be resolved and we can open ourselves to guidance from our intuition. When we can believe that we are living in freedom, joy and love, we create that level of energetic expression in our awareness, and we attract the experiences of those energies.

By examining our beliefs about ourselves, we can decide if we want to continue to live under their limitations. There is no requirement outside of ourselves that we subject ourselves to limitations. They are part of our game and the roles that we have assumed for ourselves. In order to change our roles, we need to change our vibrations and our state of being. We need to align our imagination and emotions to a new level of polarity and frequency. If we can open ourselves to our heart-consciousness, we can find that we have the perfect guidance for realizing our natural state of gratitude and joyous fulfillment. Our new level of vibration brings our life experiences into resonance with us. In this way we can realize that we have the ability to elevate the quality of our life experiences by resolving and transcending our beliefs about ourselves through our inner knowing.

Awareness Beyond Time/Space

There are many dimensions of experience available to us, that our ego-conscious self knows little, if anything, about, including our own identity and essence. Our ego-consciousness operates in the mind and is mostly unaware of our heart-consciousness and the workings of our subconscious. It is mostly through our subconscious that we may be aware of certain promptings in our inner knowing. Our subconscious participates in unlimited consciousness. It provides infinite awareness to us, but we receive only as much as our beliefs about ourselves allow. Our beliefs set the limits of what we allow ourselves to realize as reality. Without belief in personal limitations, we can be aware of our essence, which is our conscious present awareness, including our connection to universal consciousness. It includes awareness within our entire subconscious and the intelligence of our heart; the conscious connection with our guides and angels and the consciousness of the Being who constantly creates us. In the flow of conscious life force with the gift of free choice in expressing our personal energy, we create experiences that we share with all conscious beings through our awareness. The greater our awareness, the more clearly-directed we can be in our choices. The extent of our awareness is actually a choice on our part.

We are created to be creative and to create experiences within our awareness. In our current spectrum of reality, we have agreed to participate in a convincing world of empirical duality, that we have allowed to fill our awareness. To be able to open our awareness beyond this realm, we can learn to pay attention to our inner knowing. There are many ways to do this. We can breathe deeply and rhythmically while imagining that we are expanding infinitely in awareness. We can follow our natural inclinations to want to live in love, joy, abundance and freedom. These are all included in our conscious life force. We are fully provided for in every way that we could want. We just have to realize this. Since we have created subconscious limits for our awareness, we can also unlimit ourselves. If we are intent on opening our awareness to infinite consciousness, and we are receptive, the realization comes to us through our intuition. By deepening our focus, while continuing to follow our inner knowing, we can penetrate the limits we have lived within. We can be inspired to live in gratitude and joyful confidence.

It is helpful to ground our connection with Gaia. We can be alone and barefoot in nature, especially in inspiring, beautiful and majestic places. We can listen to the birds and other creatures in the forests, feel the energy of the trees and immerse ourselves in wild water. It can be helpful to listen within the silence of the high mountains and tune into the sounds and melodies of the planes of existence.

Greatness is available to all of us in our own essence. To realize this, we can open ourselves to it and invite it into our awareness. We can imagine the energetics of Beings of light and love. We are those Beings when we realize our true identity and infinite creative ability. We can transform ourselves into knowing our essence and expressing our true Self-Realization.

Facing and Resolving our Personal Challenges

We are created to be sovereign and free, eternal creators of experiences of unconditional love and joy, but we decided to limit ourselves in order to experience conditions that were unavailable to us in our true Selves. To make our experiences in empirical duality believable and meaningful, we had to blind ourselves to our true nature. We had to believe that we are mortal and fearful and can travel the path of self-diminishment and apparent self-destruction. Without knowing the process of transformation that we went through in our incarnation and that we can reverse, we have been doing our best to avoid the results of our attention and alignment with negativity. We haven't even imagined that we are the creators of all of our life experiences.

We can take our blinders off and open our awareness beyond the reality of negativity. Because we have allowed ourselves to be trained to believe in the reality of fear and death, we experience the results of those limiting beliefs in suffering and physical death. By changing our attention from negative thoughts and experiences, to positive scenarios filled with life-enhancing energies, we elevate our awareness to what we love. When we do this, we create the experiences of a greater reality beyond negativity. It can give us an understanding beyond our drama in time/space, opening ourselves to our infinite presence of awareness.

For our possible understanding of human life as one of many expressions of who we are in different energetic dimensions in our present awareness, we can use our power of intentional choice to create mastery in our realization of empirical reality. We create what we believe is real. We can intentionally change our realization, which draws our experience into resonance with our psyche. Through our realization of what is real, we can change, intensify, minimize or transform our roles as human actors.

Our Self-Realization depends upon our openness to our heart-consciousness and our willingness to transform our lives. If we are intentionally open to expanding our Self-Awareness, we can begin with imagining an experience that is a little beyond our current reality, but that we can believe is possible for us. By intentionally choosing in complete confidence to believe that it is real in our experience, we create its energetic quality, attracting resonating energetic patterns to manifest in our experience. Once this happens, we can go for something more challenging, until we absolutely know our inner guidance and our eternal present awareness beyond ego-consciousness.

Reaching for a Deeper Understanding of Life

Our culture presents God as an entity that is separate from us, if there even is a God. In experiments that anyone can duplicate, physicists have determined that every entity is conscious, based on how photons know where other photons are; how they communicate with one another and how they can become fractals of themselves in order to illuminate something and fill all pathways to their goal. These abilities indicate intelligence, intent and awareness on a subatomic level. This and other experiments with subatomic energies prompt us to conclude that all waves/particles display consciousness. They are the essence of the expression of everything. If all of our constituent entities, from sub-sub-atomic on up to everything comprising our bodies is conscious, then our bodies also must be conscious beings.

We are intimately connected with our bodies, and if we are unusually perceptive within, we can realize the intelligence of our body's constituent entities, and we can feel their subtle presence, their state of vitality and their vibratory expression. These things are present in universal consciousness, awaiting our awareness and recognition. Once we are capable of a deep awareness of our body-consciousness, we can detect the symbolism of any defects that our bodies may have. These indicate a blockage of life force as a result of limiting beliefs about ourselves, resulting in physical deterioration. We have etheric inflammation.

Physical defects have a quality of energy imbued with negativity and some level of fear that we hold about ourselves. The subject of our fear is symbolized by what our body-consciousness has created in resonance with our conscious vibratory spectrum, our energy signature.

Fear of suffering and diminishment, and ultimately fear of termination are based on our belief in our mortality and separation from God. We have been taught that we are independent entities, perhaps with a spark of life that we received from our parents and on through our ancestors to the first humans, who somehow arose and expressed conscious and separate physical life. The problem with this perspective is that physicists have found that the electro-magnetic essence of our empirical world, consists of waves-particles. These are consciously connected with all other wave-particles everywhere and know where each other is at any moment and what form they are expressing. If all sub-atomic waves/particles have at minimum a species-wide conscious awareness, they are consciously connected as well as being individuals. Separation appears physically, but the essence of every entity is part of a universal consciousness that fills everyone.

If there is only one consciousness that every entity participates in, there can be no separation of consciousness between entities in our essential being. In our deepest awareness we cannot be separate from the Creator of all. All conscious life force arises constantly within universal consciousness, imbuing all entities with conscious, individualized life force. We are the ones who can wield our conscious awareness in ways that we freely choose to express ourselves. Once we free ourselves from our belief in separation, we can realize our expanded awareness beyond ego-consciousness. We can feel our heart-consciousness and realize the reality of our inner knowing beyond the empirical world of duality.

Our Inward Journey to Infinity

At some point in our journey inward we realize that we can be however we want. When we can clear our mind and feelings and just be present in awareness, we can transcend our limiting beliefs about ourselves. Our awareness has no limits beyond our beliefs. Every moment offers us a choice of what kind of vibrations we decide to pay attention to and align with. While we are just being our present awareness, even in challenging situations, we can realize that we have an inner knowing and feeling about everything in a way that is compassionate and wise.

When we align with the energy of our heart-consciousness, we can experience everything we truly love and desire. We just have to be in mental and emotional alignment with the Source of our conscious life force and our identity. We are all the same Being in universal consciousness, and each of us is a unique expression of Creator essence, having awareness throughout the consciousness of the Creator of all. We have the ability to create expressions of ourselves, fractals of our Source, and extensions of our essence.

In multiple energetic dimensions, which are bands of frequency and polarity, we are expressing ourselves as individual beings, one of which is our physical human presence. Our physical presence is the conscious expression of our energy signature, as felt in our subconscious self. By learning to elevate our attention to positive, high-vibratory thoughts and feelings, we can become aware of our heart-consciousness and the understanding and feeling of unconditional love and joy. These naturally arise in our awareness when we summon them in a state of clarity, openness and receptivity.

Because we are fractal creations of our Source Consciousness, we are energy modulators through the qualities of our attention and emotional alignment, limited by our beliefs about ourselves. In every moment we have free choice about what we focus on and how we feel about it. By having deep desires for how we want to be, we are invited to realize the fulfillment of those desires. Since we live in a universe of energetic balance, our needs and desires must have their fulfillment to maintain energetic balance. We have the choice of being oblivious to it or realizing our fulfillment.

We live in a time of change in magnetic polarity. Even the Earth and the Sun are changing polarity. As humans learn to pay attention to scenarios that we want, we allow negativity to disappear into its non-existent essence. A shift in solar and planetary polarity would also shift human polarity. We have the choice to make the leap in conscious awareness beyond dualistic polarity into infinite presence of awareness.

Transforming Ourselves in Our Daily Lives

When we each think of ourself and how we feel about ourself, we are realizing the expressions of our energy signature. This is how we believe we are, and how we express our perspective. We can realize that every aspect of our body and personality has qualities that we can control. These qualities have energetic patterns that we can modulate as we choose through the focus of our attention and our mental and emotional abilities. By using our imagination to create wonderful and inspiring visions of inner light in every situation, we can learn to transform ourselves into the beings that we truly want to be, and our inner-light awareness becomes real.

If we find that all of our needs and wants have been fulfilled and continue to be fulfilled, we gain the freedom to live in the energies that we most love and value. We can teach ourselves how to do this by paying attention to the conscious life force that comes to us through our heart-consciousness. This can be the most energizing awareness in every moment, while we also live our lives in society. Once we decide to pay attention within ourselves to what we truly know and feel about our infinite presence of awareness, our conscious expansion can become infinite.

We know scientifically and intuitively that there is a universal consciousness that creates everything and everyone constantly in every moment. Everything in the physical world consists of swirling patterns of light-energy, appearing as subatomic entity-particles, comprising atoms and molecules and on up to the entire structure. Everything is energy that is held in form as an expression of universal consciousness. We are expressions within this constantly-creative and lovingly-directed consciousness. We are the creators of new experiences just by our nature and how we think and feel about ourselves.

Our thoughts and feelings about ourselves are the creators of our personality and physical presence. When we learn to control our attention and what we choose to be aware of, we can change our akashic history and our present circumstances by realizing the nature of our presence in every situation. We can pay attention to our heart-consciousness for moment-to-moment prompting, and we can live with confidence that we always know how best to be in every moment.

Our attention is ours to direct wherever we want. If we want to realize the spark of conscious life force in everyone we encounter, this realization will arise within us. It is present in our heart-consciousness, along with everything that we could ever truly want to experience, waiting for our realization.

Higher-Dimensional Living

We can realize our infinite awareness by feeling and knowing our heart-consciousness. This consciousness has no thoughts. It has only knowing and feeling the infinite Being of our true Self. This consciousness is much more deeply aware of us than we are of ourselves. It works with vibrations. In every moment it offers us everything we could ever want, waiting for our realization of its reality. Only our belief that complete fulfillment is impossible for us keeps us from realizing what we truly want.

When we have the desire for expansion of our conscious awareness, we are ready to take the journey to infinite Being through transcendence of our limiting beliefs about ourselves. We can recognize that they serve the purpose of intensifying our empirical experiences, because they do not allow for awareness beyond space/time. We have felt that our sense experiences are all that life can consist of, and we have confined our awareness to this spectrum of positive and negative energies.

As we open our awareness through our heart-consciousness, the first change that we notice is the better quality of life that we experience. The heart and higher chakras have no negativity. We can transcend it by directing our attention and energetic alignment to our heart-consciousness, which we can be aware of as unconditional love in the consciousness of the Creator, within which we arise as expressions of divine consciousness, having infinite creative power.

In our ego-consciousness, we cannot imagine our greatness.

The way out of limited consciousness is transcendence through intentional awareness and alignment with clarity in heart-consciousness and the higher chakras. Once we can penetrate our limiting beliefs to their origins in fear, anger and doubt, they become unbelievable for us, and we can realize our infinite presence of awareness as our essence beyond any dimensional expression of consciousness.

Although we have chosen to participate in the human game of empirical duality, we have become unnaturally fixated in it. We are the energetic expressions of Creator consciousness, unlimited in every way. In our essence, we are beings of light and unconditional love in universal consciousness. When we open ourselves to the fullness of our heart-consciousness, we glow with the light of our Creator, emitting massive quantities of photons. In this expanded awareness we can realize our fulfillment in every way, providing a higher dimension of living for us in alignment with the positive, high-vibratory energy of Creator consciousness.

The Extent of Our Consciousness

By holding the fear of life-diminishment in our awareness, we align with its polarity and feed it our creative life force. There is no requirement outside of ourselves to hold a belief in fear. It is a product of our imagination and a trick in our emotions. We have become so deeply entranced in the empirically dualistic world, that we have been unaware of our own role in our experiences.

Everything in our awareness is contained in our own consciousness. Although other beings can be playing the same matrix game with one another, our participation is important in the way that it affects our own creativity. If we are entirely motivated in alignment with the intuition of our heart-consciousness, we have our own mental and emotional guidance within ourselves. There is nothing to fear, because intuitively we know and feel what is true for us.

We can realize that we have greater consciousness beyond the realm of duality. Beyond duality is an energetic realm of life enhancement in every way. Negative energy cannot exist on its own, because it is self-destructive. It needs a host to provide conscious life force for it. Humans are the hosts for negativity. Instead of reacting to negativity in our presence, we can continue to realize our presence in deepest love and joy.

It is possible to realize ourselves beyond the physical body. We have awareness beyond time/space. We are eternally self-aware within universal consciousness. Our physical body is an expression of our subconscious self in alignment with our energy signature, our state of being. By following our intuitive knowing with gratitude, we can transform our lives into the realm of joy, compassion and fulfillment.

If we perceive a threat, it means that we have lost our intuitive connection, and we are in duality and fear, and we believe in our mortality. If we find ourselves in this situation, we are free to change our focus to what we love in order to shift our polarity to positive, resolving the negative energy. Without our fear, the negative is powerless in our personal life. Courage is required to take this leap in consciousness beyond fear to unconditional love. It's easy if we follow our intuitive knowing, even if we waver as we're learning.

With powerfully impressive intention, we can urge our subconscious to work in alignment with us, as we expand our awareness to the essence of who we are as our infinite presence of awareness with creative abilities. Our expanded awareness can include the entirety of universal consciousness. Once we can realize this, we are personally unlimited in every way.

This life is a mental and emotional game that we're playing with our consciousness, together with other humans and celestial beings. If we desire full Self-Realization beyond time/space and duality, it is all present in our consciousness already, awaiting our realization.

Deepening Our Realization

Once we realize our expansive Self and eternal presence of awareness, we do not need ever to subject ourselves to personal limitations again, although we can express our presence in the dualistic world of humanity. In this realm of consciousness, there is negativity, but we can continue to live in the consciousness of our heart, where there is only life-enhancing energy. From the limitations of ego-consciousness, this is foolish, because at this level we believe in fear and mortality. In the infinite awareness of heart-consciousness, there is only eternal presence of awareness of whatever scenarios we choose to pay attention to, including the void of all potentialities, as well as all forms and manifestations. All of this awaits our realization in order to arise in our experience.

As we become willing to be responsible for the expressions of our life, we can realize that being responsible means that we are the creators of our experiences. We are completely self-sufficient in every way. Our limiting beliefs are the only things that keep us from realizing who we are and what our capabilities are. Because we are fractals of Creator Consciousness, we have the ability to be aware of anything and anyone we choose in any circumstance. Within our awareness we have understanding and attention. Beyond personal limitation we can understand the nature of human life and the drama that we are living through.

To awaken heart-consciousness in humanity, some humans have to establish and strengthen the presence of expanded awareness and to live in its energetic spectrum. That is the reason for the large light-worker presence on this planet now. Some of us can do this, and it is supported and enhanced by the consciousness of the Spirit of the Earth and the Sun. We are all potentially like the Sun in our essence, if we choose to express ourselves as stars, which some of us may currently be in another dimension. We can choose to live in infinite awareness, while enjoying great creativity.

We actually are never given a difficult situation, unless we have aligned with its vibrations. When we are aligned with our heart-consciousness, we have only love and joy, because we know everything about our condition and understand how to control our focus of attention and emotional alignment in confidence and compassion. We can teach ourselves to see the light in every conscious being and to interact with it, while transforming any negativity that may intrude, by knowing and feeling our creative mastery of every situation. This perspective creates a powerful presence that negativity cannot exist with, and so we can experience wonderful and miraculous encounters.

Living as Our True Selves

Beyond our physical body and the empirical world, our conscious awareness is unlimited. Our essence is a unique personal expression of universal consciousness, infinite and eternal. How do we know this? If we are interested in finding out, we can study the basis of quantum physics, we can read the accounts of those who have had extensive out-of-body experiences, we can study and practice intensive meditation techniques, and we could enhance our natural pineal gland secretions of dimethyltryptamine with ayahuasca or other powerful psychotropics that stimulate our inner knowing. At some point, this inner knowing becomes our natural state of being, as we evolve through the ascension experience.

It all has to do with our desire and motivation to experience our soul-consciousness through our ability to be open and receptive to the joy and unconditional love that comes to us when we align with our intuitive knowing in the energy of our heart-consciousness. For this experience, we must transcend ego-consciousness through resolving our limiting beliefs about ourselves. We need some kind of breakthrough experience that shatters who we have believed ourselves to be.

Living in unconditional love is threatening to ego-consciousness, making it impossible for us to trust what we deeply know. As long as we believe in the reality of suffering and personal demise, we cannot live beyond the ego. We need a personal experience of transcendence, which can come only when we deeply desire to know the depth of our Being and can open ourselves completely to divine love. When we can open ourselves so deeply, we can realize that we are the Self-expression of the greatest love and joy that creates and sustains all life everywhere.

Our essence is beyond words, thoughts and concepts. We are even beyond emotions and extra-sensory perception and all forms of energetic expression. Realizing our psychic presence is a step toward knowing ourselves, but we can only know our essence directly and realize our Being intuitively in alignment with our heart-consciousness. In this state of being, we are self-sustaining and self-fulfilling in every way, filled with infinite creative power in deepest love and compassion. In full Self-Realization, we are the masters of every aspect of existence. This may be lifetimes away for most of us, but it is our potential, and some of us can realize it.

Aligning with Our Cosmic Environment

We are being invited to change our realization of the quality of our reality. The cosmos is challenging us to disconnect from negative energy, because our energetic environment is becoming increasingly positive and kind. As the resonant frequencies of the Earth rise, we must adapt to them in order to continue to live here. If we cannot become heart-conscious, our biological systems become unstable, and our lives come apart. When we align with negative energy, we become self-diminishing, and our personal environment becomes overwhelmed by the power of the increasing environmental resonance.

Our constant focus upon the energies of gratitude and joy brings us into alignment with our heart-consciousness. In this state of being, we are beyond negativity, and it does not enter our lives. While we make the transition to transcendent Being, we must deal with the depths of our consciousness from which arise negative feelings of shame, guilt, anger, victimization, fear, doubt and depression. We may be able to resolve some of these through therapy, but meanwhile they cause chaos in our lives. To resolve the root of all of them, we can recognize the nature of fear, which is based on our belief in personal diminishment and mortality.

We are not required to get old, sick and die. The experiences of them are a result of our belief in applying them to ourselves.

By refocusing our attention, we can resolve the root of these beliefs and then release them to be transcended by what we love. Our entire lives in the dualistic empirical world are confined within the energetic spectrum of ego-conscious experience. Our ego-consciousness cannot allow us to open our awareness beyond fear. To do so, we need a more powerful inner guidance source that will enable us to transcend ego-consciousness.

As the most powerful energetic source in our essence, the heart of our Being is the conveyance of conscious life-force arising in universal consciousness. It gives us as much vitality and conscious awareness as we allow ourselves with our limiting beliefs. When we resolve these beliefs, we can open ourselves to the entirety of universal consciousness, and we can realize our energy-modulating ability in constantly creating our experiences.

Because our heart resonates with infinite love and vitality, these are the vibrations that we must align with in order to live in a higher dimension of consciousness. Whatever interferes with our thoughts and feelings on this level is a personal limitation that we can resolve and transcend through our conscious intention. We have the ability to choose our state of being in every moment. The more we can pay attention to what arises within our psyche on a positive level, the closer we come to mastery of our personal world.

Transitioning with Earth-Consciousness

As long as we believe that we are subject to negativity, we give our life force to enable its existence for us. Every aspect of our lives has been diminished by the negative polarity of dualistic empiricism, always ending in self-destruction through belief in personal disintegration and mortality.

When we can realize how our nature is creative, we can direct our attention to the kind of life we love, and that is supportive of all life in every way. We are the creators of our state of being in every moment. By our choice of attention and energetic alignment, we can create experiences of gratitude, joy and love. When we live in those vibrations, we align with the creative consciousness of the Creator of all. We are expressions within universal consciousness.

We are participants in universal consciousness, although we have limited a portion of consciousness to our awareness, it is our choice to create and manage our limitations. Without them, we can realize our clear, eternal presence of infinite awareness. It is as if we are TV channel tuners that can tune ourselves to our choice of channels. Humanity is generally attuned to only one channel, but there are hundreds of other channels to attune to. These are dimensions of reality for us. Our reality is created by our realization of what we imagine we are experiencing. This is the reason the same experience can be experienced differently by everyone present. It is also the reason that using our creative visions and believing in them is using the same creative ability that arises in our reaction to physical experiences.

The energy of our galaxy and the Earth is now positive and moving into pure life-enhancement for all conscious beings. The humans who are primarily negatively-oriented are becoming unstable and insane. This is evident in our leaders, who have been parasites on us, in need of our conscious life force for their presence in our experience. Every inhabitant on this planet must adjust to positive vibrations in order to continue to live here. This process may take a few years, and it may happen very soon. There are now enough enlightened humans to shift the boundaries of the consciousness of humanity into a higher dimension of our reality.

We can have access to more dimensions in our awareness by recognizing them, realizing what they are, and using our attention to resonate with their vibrations. We can become aware of them by imagining what they feel like and being receptive to the feelings in our awareness. With our power of free choice in how we feel, we can choose our energetic alignment and expect our experiences to reflect that energy.

Conscious Human Evolution

We all have a desire for pleasure and ecstasy. What keeps us from experiencing them always? We diminish ourselves by changing our attention to negative, lower-vibrating energies that we become aware of, although we have the free choice to pay attention to any quality within universal consciousness. Our fascination with negativity has its roots in the belief that we are our physical body and its expressions. Empirical energy is very compelling to the point of fixation, creating powerful limitations to our awareness. This causes feelings of lack and isolation as an individual being. This is all unnatural for us, and is destabilizing our awareness, which feels threatening. A desire arises in us for awareness of our complete Being, so that we may know our eternal essence in the consciousness of the Creator of all.

We naturally participate in all of life everywhere and in everything. We live in an energetic system that is creative and life-enhancing. Our consciousness comes with our life force within universal consciousness. Whatever we choose to do that is different from the quality of our life force cannot sustain a vibratory pattern that does not resonate with our inner being. It becomes destabilized and dissipates. That's why we suffer, become decrepit and physically terminate.

This is all pretense. We have created our own limitations in our consciousness. To be effective, they all depend on our beliefs in their reality. By our realization of them, we recognize and come into alignment with their vibrations. If we can expand our realization to the consciousness of the heart of our Being, all negativity disappears from our personal experience, because our heart-consciousness enhances all of life always. Negativity diminishes life always.

Abundance cannot happen just for ourselves to the exclusion of others. All must be included in order for completely-fulfilling life enhancement. Realization of life-enhancement for all gives us awareness of infinite love uniting all conscious beings everywhere and always. We never have to leave this state of Being in infinite presence of awareness of everything, all in the present moment. Human life then becomes a game in our consciousness, and we can intentionally play it successfully for the benefit of all. We all have the ability to do this. It requires alignment with our heart-consciousness to guide us in infinite love and joy. All of this depends upon our intention to align with the consciousness of our Creator Self. As fractals of creator consciousness, our abilities are infinite.

Choosing Our Quality of Life

Being responsible for everything that happens in our lives as humans requires our realization that in some part of our consciousness we aligned energetically with the vibrations in our awareness, and we continue to realize them as real. We are the ones who choose what kind of vibrations we pay attention to. No one can change that. We can flow through life in resonance with our heart-consciousness. This is the source of our conscious life force always. We have the free choice to align with it or not. It is our connection to universal consciousness.

This connection opens to us when we can give up our attention to negativity, and we can focus on what is most life-enhancing for all of life. This focus opens us to awareness of infinite love in our essence. We can know that we have limited our awareness with the limiting beliefs that we accepted. in a way that is real for us, we needed to limit our awareness in order to participate fully in the human experience.

When we feel that we have experienced enough in the realm of duality, we can choose to change our attention to completely positive in every moment. This is a complete reorientation of our awareness. Guided by our heart-consciousness, we can imagine the kind of life we would most love to have. When we can feel this deeply and continuously enough, we can realize its reality for us. We can realize that we are in an energetic level that brings joy and love. By constantly realizing that we are living in gratitude and joy, we continue to create experiences of the same quality.

We have the choice of realizing anything, but to be human we have to focus our constant attention on being human. It doesn't need our full attention, especially when we realize that we have infinite awareness. We are playing a complex psychological game. Once we understand it, and become intuitively aware, we can live whatever quality of life we align with. By our constant alignment we are drawing the awareness of humanity into heart-consciousness.

Insights into the Nature of Our Reality

We are all brothers and sisters and are of the same essence, arising within universal consciousness. Every sub-atomic swirling energy-entity, atom, molecule and the entire structure of our physical presence is conscious. All, even rocks, are of the same creative essence of life, which pervades universal consciousness. This is the source of all conscious life force in existence and continues to expand as awareness expands. Our purpose here is to create experiences for universal consciousness, which is what we are constantly doing. The energetic pattern of every thought, feeling and action is forever held within universal consciousness. In our ability to choose the focus and alignment of our state of being, we create experiences, which we participate in.

Every constituent entity of our physical presence has a natural polarity and vibratory range that is aligned with our heart-consciousness. When we are in resonance with this energy, we have perfect health and well-being. To the extent that we give our attention to a different kind of energetic pattern, we disrupt our personal energy signature, creating instability in our physical presence and resulting in defects in the functioning of our bodies.

If we use our bodies as symbols of our conscious state of being, we can examine our defects for clues of the kind of energy interference we are experiencing. If we penetrate the energy of our defects down to their source, we find that we have become unable to realize the infinite love and enhancement of life that we constantly receive and that keeps us alive. Through our intuition we become aware of this, when we open our minds and imagination to it.

Once we recognize our inner knowing and realize it, we are on the path to mastery of our situation. Limiting beliefs become unbelievable and dissipate from our consciousness. Without our creative life force, which we provide through our attention, they cannot exist within universal consciousness. Only life-enhancing consciousness exists in reality. We have learned to use our creative ability in destructive ways and have experienced the results. We can shift our awareness at any time to another energetic level of expression. We can use our imagination to create form and feeling to direct us to the energies we love and want to experience.

As we come into alignment with our heart-consciousness, every aspect of our presence becomes clear and brilliant. Out of far memory, we can recognize the ability to make real whatever we imagine. We are the creators of our reality through our ability to recognize and realize energetic patterns in our imagination. We all imagine our reality, and it is within our own consciousness.

Life as a Play of Consciousness

When we create positive vibrations intentionally, we enhance our personal lives, and we align ourselves with the energies of nature. We provide a vibratory environment that enhances all of humanity. If our state of being is entirely positive and felt as gratitude, love and compassion, it is in alignment with universal consciousness through our intuition. Because we are all the same essence of consciousness, we affect everyone with our own attention and resonance. When we align with the source of our conscious life force in life-enhancing vibrations, we have the infinite power of universal consciousness guiding and advising us for the greatest expressions of infinite love and creative power. Because we are all the same Being in our essence, the energies of everyone are within us all constantly.

Everything that happens is within our own consciousness. Regardless of what energies are vying for our attention, we have the choice of directing our focus to the qualities of life that we want to experience, and to imagine and feel those qualities now. We have the power to choose the subject of our attention and our perspective about it. If we want to create abundance for ourselves, but it involves diminishing someone else, we are coming from ego-consciousness. The ego does not understand that everything that is true in our lives is supportive of everyone, including us. This is part of the magical power of unconditional love.

If we are in alignment with life-enhancing energy, and we

realize the essence of our free will, we cannot help or hinder another being unless we are invited. When invited, our true help is energetically pervasive on all levels. Everything we realize about that being is of the light. We are not required to recognize any negativity or realize that something negative could be real. Negativity is energetically repelled from us without any of our conscious life force to support it.

Being aware of the nature and operation of the electromagnetic energies that underlie our dualistic empirical world, is a first step toward enlightenment. Once we know that everything is a play of consciousness, and that we are the creators of our play, we can use our own awareness to create the lives we want and to benefit everyone as well.

Expanding Our Realization

We live in a dimension with billions of other humans and other creatures, and yet we all share the same consciousness. How is this possible, because we have believed that we are all separate beings? This belief aligns with the electromagnetic empirical world of our experience, which is entirely an energetic matrix held in our experience by our beliefs. Only when we can transcend our beliefs can we change our experience.

Through experiments that anyone can perform and obtain the same results, quantum physicists have determined that there is one universal consciousness that all sub-atomic beings are part of. Since they comprise the entire structure of our physical presence, we may deduce that we must also participate in universal consciousness, but how can we realize this? Our scientific knowledge is a first step toward realization. It can help us understand the interactions that occur between all conscious beings, and how we make our reality by our recognition of the energetic patterns that we imagine.

This awareness has been part of esoteric spiritual knowledge and practice since ancient times, and it enables us to master every life experience and participate in the universal consciousness of the Creator of all. Ancient teachings reveal that we all naturally have this awareness, but until we can transcend our ego-consciousness we do not have access to it. We have learned to live in a realm of duality, immersed in fear and negativity, requiring us to believe in entropy, disintegration and mortality. None of this is believable in our expanded awareness.

Our realization of the limits of our consciousness is possible for us as a personal choice. We have control over what is real for us. We can carefully examine our beliefs and their basis. Everything based on fear is life-diminishing and limiting. We are not required to hold these beliefs. They are a personal choice, and they are set deeply in our consciousness. If we want to resolve them and free ourselves from negativity, we have the ability in our heart-consciousness.

When we shift our awareness from negativity to heart-consciousness, we enter another dimension of energetics that is beyond negative experiences. We can intentionally open ourselves to this inspiration and invite infinite love and realization of universal consciousness to fill our awareness. As fractals within the consciousness of the Creator, we are designed to be self-sufficient and fulfilled in every way. Sharing the consciousness of infinite love and sovereignty, we can realize that we are all the same Being.

Creating a Presence of Love

We all share the desire to survive and thrive, but our ego-consciousness is based on the belief that we are separate entities and that we can suffer and die. Although out-of-body experiences have disproved the inevitability of this, holding these beliefs continues to create the apparent reality of our victimization and mortality. Perhaps we can remember that we developed fear, and we consciously or subconsciously provided the desire or antipathy to create these beliefs.

We have the choice of forming and holding beliefs about anything we can imagine, and we can learn to direct our choices to the feelings that we love. Although there is social pressure to create the same beliefs shared with everyone around us, especially those we like, what is the basis for these beliefs? Our beliefs make the empirical world our complete reality. Potentially everything else is fantasy. If we examine the physical world, and we penetrate its apparent reality down to its smallest constituent parts, we find that the atoms and their constituents are swirling patterns of electromagnetic waves. Each of them has an identity that we make real for ourselves through our recognition and realization. These collections of energetic patterns have a conscious group presence, which includes our bodies. We express our perspectives and beliefs about our personality and physical presence within the vibratory range of empirical energies.

Our technology has recognized a wealth of energetic patterns beyond our physical perception, but it has not given us their feelings about themselves. Through our intuition, we can know these feelings. Our emotions and our intuition are not controlled by our limiting beliefs. Through intuition we can be aware of realms beyond the energetically-dense realm of space/time by opening ourselves to greater love and light, aligning our attention with feelings of gratitude, joy and sovereignty.

The universe is composed of fractals of universal consciousness. This means that everything and everyone has awareness that is appropriate for them. Without our negative beliefs about ourselves, our awareness is limitless. Our essence is our personal eternal presence of awareness, within and beyond ourselves. We can open our awareness to the source of our conscious life force. Although this is in a dimension beyond words, we can know it when we are giving it our attention and feeling its vibrations.

Once we can consistently direct our attention to the expressions and manifestations of love and joy, we no longer need to pay attention to anything less desirable. Our attention and energetic alignment shape our personal energy signature, which attracts life-enhancing experiences for us and radiates our creative energies through our aura and beyond.

Knowing beyond Our Human Ability

When we are in alignment with our heart-consciousness, we can use our imagination in creative and life-transforming ways. We can imagine that we are the Creator, infinite in every way and present in full awareness of all potentialities everywhere and always. We can choose how we feel about ourself in the most wonderful and magnificent ways. If we want to experience ecstasy and love forever, we can create the presence of other conscious, infinite Beings that allow us to interact in enjoyable life-enhancing ways. We can create the kind of Beings that we want to interact and be with. Now imagine that we are those Beings.

Imagine that we are created to share in the living experiences of the Creator of all and to use our creative ability to design entertaining experiences to participate in with the Creator. Suppose this is what we are doing now. We are constantly creating experiences and interactions in our thoughts, feelings and actions. In our perspective, we get to determine the quality of our experiences, and we offer all of them to the Creator, whose consciousness we share in every moment. We have ultimate freedom to think and feel however we choose in every situation.

Whatever situation we are currently experiencing is present for us because of our design and intent. Due to the constraints we have placed upon our consciousness in order to participate fully in the human experiment, we may not realize what our current life signifies for us, but with sufficient introspection and inner awareness, it can become clear for us. All of our experiences are meaningful on some level of vibration. We are finding out what the range of negativity feels like, and we are learning how to conduct ourselves and be aware of the quality of our own state of being in interacting with negativity.

If we are perceptive and open to receiving inspiration, we can know that we participate in a greater consciousness than our biological ability would allow. This awareness comes through our own inner knowing beyond ego-consciousness. When we can resolve our fears and limiting beliefs about ourselves, we can open our awareness to our heart-consciousness and the realization of our infinite essence. We are the companions of the infinite Creator and fractals of universal consciousness.

Expanding Awareness of Greater Consciousness

As humanity advances technologically, we are becoming aware of electromagnetic frequencies extending into millions of cycles per second, far beyond the receptivity of our physical senses. By our awareness, this energy becomes real in our experience. We can assume that there are even greater frequencies beyond our current ability to measure. Perhaps we wonder what vibrations of trillions of cycles per second are like. The cycles of our conscious life force are even faster than that, and out to infinity. In the depth of our consciousness, we are aware of these vibrations. They exist beyond our limiting beliefs about ourselves, by which we bring our awareness down from infinity.

If we choose to transcend our limiting beliefs, and we are open to resolving them through our realization of their basis in fear, we can open ourselves to Self-Realization. We can know ourselves as our personal presence of awareness, beyond the physical world and into the eternal present moment.

As we expand our awareness through technology, we expand our realization beyond our former limitations. As an extension of human imagination, technology has inherent limits of its own, but it weakens our current belief in self-limitation and may lead us to wonder about infinity as a possibility for us. Our heart-consciousness knows about this. In our essence we are our eternal presence of awareness with freedom to create whatever we want to experience.

With our imaginative intention, we create our bodies and our experiences in the world of humanity. We are constantly interacting with energies that we aren't aware of, but hidden aspects of our consciousness are aware of them, and they contribute to our creations. We can open our awareness to these subtle energies by imagining what higher vibrations feel like. When we become ecstatic enough, their reality comes into our awareness.

Always we are interacting with energies, which are expressions of consciousness. Our consciousness is our creative ability, and we utilize it through our attention and mental and emotional alignment, filtered through our beliefs. When we can be present in compassion, gratitude and joy, we can be aware of our heart-consciousness. As we live in these positive, life-enhancing energies, we transform our lives by realizing expanding awareness within infinite consciousness.

Mastering Personal Transcendence

The depth of consciousness we are aware of is a function of our alignment with heart-consciousness. When we align ourselves with negativity, we are energetically weak, because we disable our creative inspiration with doubt and life-diminishing imaginings. When we intentionally hold our vibrations high, we can transcend doubt by opening ourselves to confidence in our intuitive guidance. This is our connection with universal consciousness and the source of our conscious life force. It is an expression of our heart-consciousness. We have great intelligence beyond ego-consciousness.

Once we are aware of, and in alignment with our intuition in every moment, we can have absolute confidence in the thankful and joyous feelings and thoughts that arise in us without any outer stimulation. If we change our attention to a negative vibration, we slip into ego-consciousness in the realm of duality, which is where most of humanity is. Our heart-consciousness is beyond ego-consciousness and arises from our infinite essence.

We are on the path toward full Self-Realization. For the most powerful experiences, we designed our human person to be unaware of our real essence. We needed to create our ego-consciousness to be able to navigate the world of duality without higher guidance. Because we could not know our true abilities, we developed limiting beliefs about ourselves. We could not trust ourselves to be true always. The empirical world of duality is very convincingly real for us, and it is a challenge to imagine that it arises from a limited aspect of consciousness which we direct with our attention and energetic alignment.

With our own mental and emotional abilities directed by our intention, we can create whatever quality of experience and whatever forms we design, when we are in alignment with the energy of our heart. This is the pure energy of unconditional love and joy. It is on a vibratory level with the emotions of the soul, our higher Being. When we are aware of these feelings, we have the choice of aligning mentally and emotionally with them.

Opening ourselves to our deepest inner awareness requires practice, or perhaps a drastic experience of transformation. Once our life choices carry our awareness into universal consciousness, we can easily master this game of consciousness that humans are playing with one another. In our return to the consciousness of the Creator, we become our infinite, transcendently-radiant Selves.

Our Participation in Universal Consciousness

If we can imagine the idea of universal consciousness, we may wonder what it feels like. It is something that becomes known for us as we are able to align with its vibratory presence. Every part of it contributes to the enhancement of the whole and all aspects of everything. The anomaly is the dark force. Negativity does not exist in universal consciousness, but it exists in human experience, because we provide its reality for us by imagining it, recognizing it and realizing what it is. When we change our realization, negativity no longer exists in our experience.

Making something real is what we do as fractals of universal consciousness. If we are attracted to negativity, we create its presence in our awareness. This happens most powerfully in our subconscious self, which we can be aware of and communicate with consciously. This kind of communication happens as subtle feelings from deep within us, which we send and receive. We cannot make them go away, but we can block them from our awareness by turning our attention to the control of our ego-consciousness. This aspect of ourselves consists of all of our limiting beliefs about ourselves, and it is our presence in the human world. It does not know about higher guidance.

In order to open our realization beyond our fixation to the material world, we must be willing to transcend our limiting beliefs about ourselves. This involves a complete transformation of our lives. Our desires change to the life-enhancing desires of our heart. Our needs disappear, because they are naturally fulfilled as they arise. In every moment we can do what we know in our heart we are guided to do to create the most love and beauty within and all around.

Being aware of universal consciousness can happen when we are completely aligned with our heart-consciousness. This is how we can open ourselves to the most wonderful, life-enhancing experiences. When we can feel and realize the energy of the heart of our Being, we feel grateful and ecstatic. It feels so wonderful! By being aware consciously of these vibrations as the essence of our true Self, we can realize our infinite presence of awareness arising within universal consciousness.

Guided by our heart-consciousness, we can live in the awareness of our essence merging with the essence of everyone and everything in the same Being, within whose consciousness we all live. In this perspective we are telepathic, empathic, telekinetic and every unlimited form of communication and knowing. This all happens when we open ourselves to participation in universal consciousness.

Aligning with our Essence

Although we have incarnated here without understanding what we intend to accomplish in this lifetime, and how we can create the experiences we truly want, we can become aware of these things and much more. We face great challenges by being programmed with false information and limiting beliefs, and we take on the fears, sense of separation and lack that imbue our society. We have not learned to develop sensitivity to our own inner knowing, which holds the fulfillment of everything we could ever want. Once we catch a hint of our potential, we can transform our lives.

We've spent eons contributing to universal consciousness in our experiences with negativity, and we've become so entranced with it that we've convinced ourselves that we are our ego-consciousness living in our physical bodies, often living in poverty, suffering the dictates of tyrants and destined to die alone. In our true Self, none of this would be possible. We make it possible by our limiting beliefs about ourselves.

Although we may believe that we can be hurt, subjugated or inspired by others, our first realization of remembering who we are is awareness of our free will in choosing how we want to feel about anything. No force outside of ourselves can dictate how we feel. Although we may relinquish our personal power, we can control our attention and how we feel about what we focus on. Once we realize this, we can begin to direct our thoughts and emotions in creative ways.

Many of us have been unaware or have been unable to grasp the understanding of quantum physics that there is nothing solid about our empirical world. Everything consists of conscious, swirling patterns of energy, beginning with the smallest sub-atomic entities. We are constantly interacting with energetic expressions of consciousness, and this includes our bodies. Our feelings and beliefs are energetic expressions of our conscious choices, including our subconscious selves, which we can become aware of and draw into mental and emotional alignment with our presence of awareness.

Because the empirical world is an electromagnetic energetic expression of consciousness, it is through conscious processes that we interact with this world. The quantum sciences have determined that our conscious recognition of energetic patterns brings them into our awareness as real for us. Apart from our recognition, they are not real in our experience. They are unmanifested energies outside of our personal experience. Every possible form and substance exists in the quantum field, expressed in consciousness and available for our recognition.

All conscious beings share the same consciousness. We are all of the same essence within unified consciousness, and this consciousness includes everything about us, our feelings, thoughts and awareness. Although we have our personal awareness, we are not separate beings with separate consciousness. We are constantly creating the experience of separateness by our limiting beliefs about ourselves. It is a play in consciousness, expressed energetically as empirical reality.

Once we become aware of our free-will choices about ourselves, we can learn to direct our thoughts and feelings in alignment with the qualities that we desire. If we open our awareness and are receptive to the feelings inherent in our essence, we can transform our lives in alignment with the consciousness of the Creator of all, within which we live and have our identity. We can direct our attention to our Source Consciousness in the energy of our heart. This is our creative essence of unconditional love and enhancement of all life, constantly interacting in union with all conscious beings everywhere.

Realizing Unconditional Love

Consciousness is the prime essence of everything, and everything arises from consciousness. It is universal. Quantum Physics has shown that there is only one consciousness, and within it arise an infinite number of fractals that all share the same consciousness. Each of us is one of these fractals. Each subatomic entity is one of these fractals. Every part of us is an aspect of the whole. Every part of each human has the same DNA. We are unified beings in a unified universe. All are held within the same consciousness. This is the consciousness of a supreme Being, who expresses us to realize infinite Self.

We, in turn, express ourselves as humans, forming a species consciousness, within which we manifest our personality and physical presence. We have access to universal consciousness, just as every cell of our body has access to the consciousness of our whole person. This unified consciousness has a magnetic attraction for every molecule and cell of our body, as well as everything that we believe belongs to us. We extend our conscious presence into the empirical world as far as we believe that we can. If we believe and realize that we experience only loving relationships, they fill our lives with their quality.

We have different ideas of what love is, but it has an essence that we can know and feel. It is the creative essence of universal consciousness. It expands and enhances life everywhere and always. It is infinite in its presence and pervades our awareness beyond ego-consciousness. Infinite love expresses itself in positive, high-frequency vibrations beyond our comprehension, filling us with ecstasy and wisdom beyond thought. It is our essential Being, which we can know and feel beyond all personal limitations, and it pervades our eternal presence of awareness and gives us fulfillment of every desire of our heart.

Through our heart-consciousness, we can fulfill our destiny by creating experiences that enhance human consciousness and expand universal consciousness. By being aware and in alignment with the vibrations of our heart, we express unconditional love as a radiance from our energy signature. It imbues our imagination and emotions with its presence.

Our participation in this energetic field is voluntary and intentional. Because humans currently do not realize the power of love, those who choose to live in alignment with its vibrations in every moment must have strong intention and deep understanding, but once we begin to live in its vibrations, we feel and know that this is our natural state of being, and it feels so good!

Being in Time and Out of Time

We live in the conscious life-stream of the infinite One. It fills our Being with divine light, infinite love and joy. We participate in this consciousness as much as we allow ourselves, having the free will to limit ourselves, but we cannot diminish our connection with the consciousness of the infinite One. It arises in us through our heart-consciousness, even in our physical bodies. Our heart loves us unconditionally and lives to enliven us, regardless of what we do to it. If we can love our heart the way it loves us, we can begin to understand infinite love.

We are constantly being created in the awareness of infinite consciousness, providing the essence of our eternal Being. We are all the same Being, sharing the same consciousness and aligning with and directing the imagination and emotions of the infinite One within ourselves. We are the directors of our human experiences, the condition of our bodies and our feelings about ourselves, and we are much more. We are infinite in our consciousness, but we are aware of only as much as we allow ourselves with our limiting beliefs about ourselves.

Once we become aware of our potential, we can begin to penetrate the essence of who we are. Beyond the body and the empirical world, we are our personal timeless presence of awareness. This awareness arises in us when we are clear and serene. It happens momentarily in our personal vibrations between cycles as the amplitude shifts between increasing and decreasing trillions of times a second. It is at the point of change, when, for a moment, there is no movement. It is a moment in eternity, and it is the moment our intuition can guide us to, if we are open and receptive to it. We accomplish this by feeling and imagining the most wonderful and joyful scenarios and feelings. These are the vibrations of our heart, filled with compassion and love. These vibrations align us with our heart-consciousness and our intuitive knowing and feeling.

When we open ourselves to positive, high-vibratory thoughts and feelings, we are beyond ego-consciousness, which knows only the limitations we have believed about ourselves. By aligning ourselves with the energy of our heart-consciousness, we can transcend limitations, as our awareness expands into unlimited consciousness. In this state of Being, we can understand our lives in every aspect and in other dimensions.

We are living in a time-wave that spans thousands of years. This wave has exactly bottomed in 2012 and is now beginning to rise. We are at the point out of time, when everything changes, and we become aware of our majesty and sovereignty. As our collective consciousness expands in positivity and higher vibrations, we create beauty and life-enhancement. This happens when we share our light with one another through our eyes and our radiance of love, regardless of whom we encounter. As fractals of the infinite One, we are able to express the unlimited creative power of our essence.

Awareness of Greater Consciousness

As telepaths know, universal consciousness enables us to be aware of the awareness of other conscious beings. Our universe is loaded with magnificent conscious beings, including the planets and stars. We are all radiant energy beings interacting with one another. Each of us determines the quality of our interactions in how we choose to feel at any moment. Our ego-consciousness presents challenges to us, all based on lack and fear, but it is possible for us to choose to receive everything while being aware of being our present awareness and acting in ways that benefit all life.

In the essence of our heart we constantly receive the conscious life force of infinite consciousness, enlivening us to create experiences, whose energy is transmitted throughout consciousness, contributing to infinite expansion. The Spirit of our planet also expresses herself through her energetic radiance, which we interact with constantly. Its energetic expression is most powerful for our awareness when we have skin touching the unpolluted Earth.

Our Sun magnetically attracts and directs the energy in our Solar System. As the clearest and brightest conscious Being in our human experience, he illuminates our lives in every way. Our planet and our Star live in the same energetic dimension as our heart-consciousness. We can receive their conscious expressions by opening ourselves to their energy and desiring to feel it and know it. This happens when we are in a state of life-enhancing feelings and thoughts, appreciating their essence and expressions.

Like our heart, our planet is unconditionally-accepting and forgiving of humanity. She continues to provide every sustenance for us, regardless of what we do to her. When we are in alignment with her energy, it is the same as our alignment with our heart-consciousness. This is our connection with our own essence and the opening of our awareness to greater realization.

We can realize that the empirical world consists of electrical and magnetic energy, just like our thoughts and emotions. As we have control of our thoughts and emotions, we can control and direct the empirical world of our own experience. We have the personal choice of creating a world of beauty and love for our experience and all of the others who make the same choice.

In our quest to be more acutely aware of our intuition, the true essence of nature, as an expression of universal consciousness, provides wonderful guidance for us to align with.

Aligning with Our Greater Self

As fractals of universal consciousness, we have access to every thought and emotion that has ever occurred everywhere, but we have imposed upon ourselves the limited awareness of the dualistic, empirical spectrum of vibrations. This has been our agreement in order to play our roles in the human experience. We are such good actors and actresses, that we have become entranced in our roles. By our imagination, we are held in the vibratory range of the world that is real for us. This happens in our realization of what we recognize and resonate with, or what we resist, but also align with.

Humans are greatly conflicted. Because we are being guided to recognize a realm beyond our senses, we have lived in a world that is uncomfortable for us on many levels. Before we experience the material world, its essence arises in our consciousness and takes form. Although our consciousness is unlimited, our awareness of it is contained within our limiting beliefs about ourselves and our abilities. These limitations hold our attention, until we intentionally change it. When we choose to open ourselves to awareness of a more positive environment, we no longer have a need for limitations.

If we choose to know our greater Self, we can open our awareness to heart-felt vibrations and pay attention to our inner knowing. Everything in our experience is symbolic in providing guidance for us to elevate our awareness. The energetic quality of creator consciousness is the enhancement of life everywhere, whereas negative consciousness is the diminution of life everywhere. When our imagination and emotions are dominated by fear and a sense of lack, our experiences involve obstacles for us to continue as we have. The energies of life are always positive and light-filled, making our experiences smooth and easy. This is our natural state, enabling us to express our greatest heart-felt passions without interference.

In order to know creator consciousness, we must align with its energetics. By being always aligned with the energy of our heart, we transform our experiences from difficult and uncomfortable to graceful and joyous. When we can attain these states intentionally within our own being, regardless of the energies we have allowed to fill our awareness, we enhance our roles in the human drama and free ourselves from our limiting beliefs about ourselves.

Living in the Flow of Life-Enhancing Energies

As fractals of the Infinite One Consciousness, we have infinite abilities, and we can access them when we become absolutely trustworthy to ourselves. We have a deep connection to the Source of our consciousness and our awareness. This inherent awareness arises from the expressions of our heart-consciousness. It is always present, and we receive as much of it as we allow ourselves.

Because the ego operates on an electric charge, whereas, the heart energy is magnetic, ego-consciousness does not recognize the reality of heart-consciousness. They work together in their essence, and heart-energy is much more powerful, but unintrusive. In order to keep our awareness away from the expressions of our heart, our mind keeps us distracted and entranced without the resonance of the heart. This is an intentional self-limitation that we impose on ourselves. The deeper we go in ego-consciousness, the more our heart is silenced by us.

In ego-consciousness, we align our vibrations with a spectrum of energy different from heart-consciousness. Comprising our ego-consciousness are our limitations of every kind. They are all self-imposed to enhance the reality of our human experiences. These are powerful teachers beyond the limitations of duality and negativity, because they express the vibratory qualities of our personal state of being and energetic signature.

If we believe that we are perfect fractals of creator consciousness, which is what we are in our essence, we can realize the power and quality of greatest love, because we deeply know that we are our present awareness beyond space/time. We can modulate the energetics of any situation to align with the expressions of our heart-consciousness. When we realize our ability, we have the power of instant manifestation of anything our heart desires and guides us to.

By being open to our inner guidance, we can begin to recognize where it is taking us. We can intentionally go there in our feelings and imaginings. When we pay attention within, they will express for us the knowing of our inner guidance. We can know it by its vibratory resonance, which is creatively life-enhancing in every way. By aligning with this quality, we receive the clarity of greater awareness, enabling us to release our limiting beliefs about ourselves and realize our essence arising within the consciousness of the infinite One. This realization is beyond words and thoughts, but we can feel it and know it.

By following our deepest inner feeling and knowing, we can create a state of being that aligns with our heart-consciousness. This transforms our human life into experiences that we love and are grateful for. We can enter the awareness of the eternal radiance of our heart and clarity of our mind.

Creating a Viable Perspective

The ego-conscious mind is laden with desires for thriving and fears about surviving, not knowing that we are powerful creators of whatever we focus our attention on and align our vibrations with. Our capacity for realization either constrains us within self-imposed limitations or frees us in awareness of infinite consciousness.

Negativity in our conscious and subconscious awareness is constraining. It is held in secret, existing deeply in the subconscious and causing experiences that displease us. By exposing ourselves to nearly continuous propaganda, much subconscious sinister programming passes unaware and without resistance to the subconscious mind and influences the vibratory quality of our state of being. We can feel this. When we feel the presence of negativity, we are subject to it, until we can control our vibratory quality.

Through expanding realization, we can resolve the emotional knots that still exist within. We may realize that human life is a play of creating believable kinds of experiences we could have only in this dimension of energetics. In our true Self, these would be just imaginary. They would not have the same impact. We have had to endure ego-consciousness with its lack of higher guidance, even though that guidance is always present in every aspect of our lives. This is a potential learning experience, prompting us to realize that we have deviated from our natural energies and need to change our focus to what feels good in our heart. This brings relief and support in every way.

We've become so entranced in our roles in the human play, and in our awareness of the happenings around us and around the world, that we have lost awareness of what is really happening. By what we realize as reality, we relate to the light or to the darkness in everyone. When we've experienced enough negativity, we can decide to change our polarity to align with our heart-consciousness.

We can direct our awareness into a state of transcendence by inviting deep compassion and joy into our personal state of Being and into our interactions. This transforms our personal lives into experiences of life-enhancement and expands our radiance to all conscious beings. We can be confident in our creative ability. By our state of Being, including our entire conscious presence, we attract experiences and encounters that resonate with us. Everything we experience is a reflection of our own vibratory presence. As we direct and manage our attention and emotional alignment, we have the choice of being constricted or being expansive, and we can make either one real in our experience.

Creating a Wonderful Life

We all participate in the infinite presence of awareness that is the essence of our Being. We can begin to imagine what this is when we can stimulate ourselves toward a state of joy and ecstasy. These are the vibrations of our essence, coming to us through the heart of our Being in every moment. If we can open ourselves to this vibratory level and align with it through our desire and intention, we can feel our true essence in wondrous ways. Our feelings in this energetic spectrum can open expanded awareness for us of infinite consciousness. In our true essence we are points of light and centers of awareness beyond the physical world and encompassing all of it and much more.

Once we realize this, our perspective changes. We have our physical presence, which we can change to our liking. Everything about us is an energetic expression of our conscious and subconscious feelings and thoughts. The radiance of our polarity and frequencies attract experiences that resonate with us. We are psychic magnets for experiences. Once we align with a new quality of consciousness, we adjust to it and continue to attract experiences on its level. This is a life-transforming kind of development.

The more we align with our essential Self, the more natural we feel, and the more enjoyable our lives become. If we ask within to be able to feel our true essence, and we are willing to align ourselves with its energetics as we feel them, apart from what we think about them, we will open our awareness to more knowing and more intense experiences of the heart. Once we know this kind of state of Being, we can return to it by aligning ourselves with the life-enhancing feelings of our heart-consciousness.

As we go through our lives, and we gain more awareness of everything, we may learn to understand the nature of the roles we are playing in the human drama. Although we've inherited a magnetic attraction for past-life energetic returns, we can rewrite the script of our roles from a transcendent state of Being.

We can enter a heart-conscious realm within, guiding us to resolve all of our personal dramas and to release them. We access this realm through our mental and emotional alignment with it, as we pay attention to its elevating expressions, which are always present for us. They are life-enhancing in every way. Through our creative ability, they inspire beautiful music and poetry, inventiveness and personal fulfillment, all resulting from the joy and elegance of our natural state of Being, which we share with all of nature and the cosmic presence of universal consciousness.

Understanding Our Potential as Creators

Let's begin with our consciousness. From quantum physics experiments, we know that consciousness is universal. It is everywhere and always and is the creative essence of everything. In our essence we arise within universal consciousness. It is the essence of our life and our awareness. Although it is beyond explanation, we know what it is. We are all fractals of it, sharing its essence within and beyond time/space. In this incarnation we are expressing our consciousness as our personal identity and physical presence, but this is only one of possibly an infinite number of expressions.

From our observation of the workings of nature, we may understand how all life forms are constantly renewing themselves, and life keeps expanding. As expressions of Creator consciousness, we have the freedom to express ourselves however we desire, and in our essence we have access to the infinite creative power of universal consciousness. Just by our being present, we are creating patterns of energy by our thoughts and emotions. We can choose their quality, their polarity and vibratory frequency. These energetic patterns create an electromagnetic vortex around us and attract resonating experiences.

In this way we create experiences for ourselves. Every experience we've ever had was formed by the vibratory qualities of our personal energetics. We have chosen our current situation, or we wouldn't be here, because in our essence we are sovereign Beings. We each have our personal presence of awareness with the choice of expressing ourselves however we want. How we feel about ourselves in any moment, whether it is in reaction to something or initiating a positive or negative feeling, affects our energetic signature and attracts compatible energies.

If we can align our thoughts and feelings with the life-enhancing vibrations of nature as expressed by the Spirit of the Earth, we can create life-enhancing scenarios in our imagination and can open ourselves to our heart-consciousness in gratitude and joy. We can open our awareness to the loving light in every person and circumstance and interact in compassion and acceptance. This state of being attracts experiences that are positive and heart-felt.

As expressions of universal consciousness, we have the ability to create any quality of experience. As we become mentally and emotionally clear and aligned with our heart-consciousness, we create beauty and joy throughout our experiences and encounters. We can create whatever we imagine enhances and expands all life everywhere. We are eternal and infinite in our presence of awareness within universal consciousness, sharing infinite love with all conscious beings.

Our School of Enlightenment

Every experience of our lives begins with a choice of energetic alignment. Since we live in a world of duality, of positivity and negativity, nothing is better or worse or good or bad. Every choice has a polarizing quality and a vibratory frequency. It is these energetics that radiate throughout our consciousness and attract experiences that resonate with us. We live in the energetic plasma that we create through our state of being. Nothing happens by accident. All of our experiences are attracted to us by our own mental and emotional perspectives and feelings in every moment.

When we suffer, feel depressed, violated or subject to an accident, it is all a result of our own energy. We are in school to learn to direct our thoughts and emotions to the experiences that we truly want. We cannot escape from the results and reflections of our own energy. There is no punishment or reward in nature. There are only events drawn into our experience by our own energetics. None of us is unintentionally enslaved or thrown into poverty or in need of being saved by someone else.

All of us are masters of consciousness in our essence. We are expressing ourselves in our human lives for our greater experience in encountering a diverse set of energies in an environment that is possible only within the consciousness limits that we have designated and subjected ourselves to through our limiting beliefs about ourselves.

Whenever we are ready, we can change our experiences by

changing our personal orientation to life. We can play with our energetics to determine for ourselves what happens as a result of our habitual thoughts and feelings. Although we have learned to react to events and encounters outside of our bodies, no one can require us to react in any specific way. We have absolute freedom to think and feel however we want at all times. Instead of anger, we can choose compassion. Instead of jealousy, we can choose gratitude. Instead of shame and poverty, we can choose joy and abundance. It all depends on our attitude. Every negative has an equal and opposite positive. We are free to choose either one and find out what happens as a result.

This school of ours has teachers, guides and angels to assist us. We can receive their help by asking for it in heart-felt ways in which we lay ourselves open to their guidance and love, and by imagining what their presence feels like and then aligning our own feelings with them. This opens us to our intuitive knowing, providing us with everything we need in every situation.

Realizing a New Dimension of Living

Although the empirical spectrum of electromagnetic wave patterns is all that humanity recognizes as the world of our reality, it is enveloped in a sea of waves and patterns of infinite expressions, all formed in the consciousness that we participate in. This consciousness envelopes everything everywhere and consists of constant creation of fractals of Self in every possible form of expression, constantly evolving into greater enhancement of all. Everything we love is getting better, and the reverse is also true. The polarity comprising our world of duality is separating into disparate dimensions, causing our way of life to change dramatically.

All aspects of our lives that have been affected by negativity are dissipating out of our experience. As the larger realm that we live in rises in positivity, we also are being drawn into awareness of the life wave flowing through our heart. It is our choice of whether we want to have this awareness. If we believe deep down in our being that we cannot have it, we must live within this limitation. But if we can open ourselves to awareness of the inspiration and infinite love of our heart, we can be in alignment with true Creator energy. This is a state of being that stimulates us to modulate unlimited life-enhancing experiences through our imagination and emotions.

If we can realize a new and more joyful reality arising in our inner knowing, we can think about and feel it in our imagination as if we are experiencing it. By being grateful for everything we experience, we align ourselves with our heart energy and open our connection with universal consciousness.

At this point, we can no longer be connected with any kind of negativity. It all destabilizes and dissolves in the presence of primal life force enhancing all life. This is what's happening in our society. Those who have held power over others are becoming insane, and their corporations and governments are breaking down. They are departing from our experience. The banking system has controlled humanity through governments and militaries, and they are now in crisis as they dissolve, while our awareness shifts to systems that serve humanity and all life. They already exist. Our part is to recognize their vibrations and align with them. As we live in these expanding energetics, our radiance attracts the awareness of others, until we all realize a new reality that is beautiful, joyful and infinitely loving.

Removing Fear from Our Lives

In our expanded Self, we do not know fear. It is an unknown vibration. We do not know doubt, depression or suffering or any other negative state of being. In order to know these vibrations, we created a realistic realm of consciousness limitations, and we convinced ourselves that we are empirically mortal. Realizing mortality is the basis of the human belief in self-limitation. Part of this belief in mortality is the belief that we are our empirical bodies. These beliefs do not allow us to realize our essence beyond time/space.

Living with us, there have always been spiritual masters and shamans, who are aware in realms beyond the physical world. Our ego-consciousness is not especially interested in those realms. It is entranced in the material world of duality. We've been able to expose ourselves to the results of negative thoughts and feelings, and now we know more than we did prior to our incarnation. By transcending ego-consciousness, we can open our awareness to our true essence of being. We can realize ourselves in our eternal presence of infinite awareness, able to express ourselves energetically as anyone we desire to be in any dimension. We can open our awareness to the awareness of all conscious beings. In our essence we are empathic, telepathic and much more. Our consciousness is infinite, and we arise as fractals of universal consciousness, able to create qualities of experiences in all realms.

This awareness allows us to live here among our brothers

and sisters of all humanity, creating and enhancing everything that we love and every experience in alignment with our heart. We now know what negative energy feels like, and we can return to our expanded Self-Realization by filling our awareness with gratitude, infinite love, compassion and joy. In this way we transform our lives with our unlimited Self-Realization, which now includes knowing fear. We can release fear by aligning our attention with the inspiration of our heart in every moment.

By knowing the consciousness of our heart and aligning with it intuitively, we can change the experience of our reality. We have an inner knowing that is natural for us and that connects us with universal consciousness. It is infinite and all-knowing of everything that everyone has ever known. It is beyond polarity and is entirely life-enhancing in the way we describe as unconditional love, and it is much deeper than any description we can provide. When we can realize its reality, it is present for us always, making all of our limiting beliefs unbelievable.

Our Higher-Dimensional Essence

We are entering an era of knowing, an era of fulfillment in every way. Arising from the consciousness of our heart, we are being attracted to intuitive knowing beyond our thinking and ego-consciousness. It is our heart-felt essence of being, and it is the same presence of awareness as every conscious being. We, however, have been trained to believe that we don't have this kind of reality. Ours has been dualistically polarized in the empirical spectrum of energetics.

Ego-consciousness does not know where our consciousness arises from or even how we know what we are. These things can be known intuitively, and this knowing is beyond time/space. When we resolve our belief in personal limitations, we realize that we are part of infinite awareness of Being with the ability to create experiences of whatever we desire with our mental and emotional expressions. In fact, that is what we're doing right now, whether we realize it or not. This is an ability that is beyond realization. It is our essence. It is what awareness and realization come from. Our essence is the essence of creation.

Because we have not known or have not been open to knowing or desiring to know these things, we have been creating our experiences randomly, without the higher guidance that is present within. Our minds and emotions have been reacting to qualities of experiences, and in the process, creating more of the same vibratory quality.

The realm of duality, which every human participates in, is a realm created by human consciousness in telepathic unity. Without our recognition and energetic alignment, we could not participate in it, and it could not exist for us. With our realization, which we can direct, we can change human experience. Using our powers of visualization and emotional projection, we can change the vibrations in our own state of being and project our energetics into the quantum field of consciousness that envelops us for manifestation into our experiences and the experiences of our race.

By becoming more aligned with the energies in the heart of our Being, we can relax into a life of love and joy. We can always know everything relevant about every situation we're involved in, and we can participate with compassion and life-enhancement, creating joy everywhere in our presence. We are evolving into complete fulfillment of our lives and awareness of our infinite potential.

Opening Our Awareness to Personal Regeneration

Our physical body arises from our consciousness, and we created the realization of our body. With our personal energetic expressions, we have attracted every swirling vortex of light comprising our physical presence, from the minutest subatomic entities, all vibrating in resonance with our energy signature. We can feel this, and our senses are aware of their empirical stimulation. Our bodies are patterns of energy held in form by our awareness and recognition. By recognizing energetic patterns, we bring them into our realization as real. If we do not recognize them, they do not come into our personal experience. If we recognize them, and we also admire them and align with them, we bring them into our space/time experience; otherwise, they're passing waves in a spectrum of energy that exists in a dimension other than ours. This is true for the qualities of our body, as well as the defects we impose on ourselves by our limiting beliefs.

When we harbor any kind of negativity, we interfere and destabilize the life processes in our psyche and our body. We focus on disharmonious wave patterns. If we want to create resonance within our personal awareness, we must align with our true vibratory essence, as expressed through the heart of our Being. This results in clearing out of our awareness the defects and life-diminishing thoughts and feelings arising from our limiting beliefs about ourselves. Depending on how closely we resonate with our true essence, we transform our physical presence in alignment with our consciousness.

This kind of transformation has often taken years to complete, but it could happen instantly, if we can completely direct our attention to the spectrum of infinite love, joy and fulfillment in every way. We can use our imagination and call on our guides and angels to inspire us. We are all aspects of the same Being and can have psychic effects on one another. Our energy is most powerful when we are in alignment with the qualities of our guides and angels, as well as our heart. When we withdraw our attention from all of our old patterns of thought and emotion, we can open our awareness to a realm of love and beauty.

As we allow our imagination to drift and our emotions to arise naturally with the intent of realizing life-enhancing experiences for all, we can align ourselves with our true presence of awareness and our intuitive knowing of the essence of our infinite consciousness.

Realizing Our Inner Light

We are being invited to be aware of our inner light. Every component of our bodies is a light-being, expressing itself as a swirling form of light-energy. Although it appears to us as physical, it vibrates beyond our perceptual limit, yet we realize it as empirically real for us. Our Sun vibrates at such intensity that we cannot look at it. Much more and we could be blinded. It would be unseen by us, but for our inner awareness, it would have a strong presence. We also have the same kind of presence in our essence beyond space/time and can know one another on an intuitive level. This presence is what we could describe as our light-being, which expresses itself as us beyond our ego-consciousness.

Our inner light is beyond the energetic spectrum of time/space. It is the essence of who we are, and it is the source of our conscious awareness. It has the quality of unconditional love and joy. It is the creative life force constantly arising within the consciousness of the One Being that we all participate in.

In order to open ourselves to this awareness and be receptive to it, we can be meditative and reflective. We are the ones who direct the quality of everything in our lives, from the intricacies of our bodies to our encounters and experiences. Everything begins with our entire personal consciousness (conscious and sub-conscious) and how we feel about ourselves most deeply. This establishes our energy signature, which is our electromagnetic expression. It attracts experiences that resonate with it.

As we are able to align with love and joy in our imagination and emotions as a constant state of being, we create the energy patterns that attract compatible experiences. This is possible for all of us, once we realize how life works, and we develop a constant focus on the kinds of energies we love the most.

In our deepest Being we are all of the same essence, which gives us infinite creative power for the enhancement of all life. When we learn to align with this energy, our outer experiences reflect the same qualities as our state of Being. There is no chaos, threat or fear. There is only the energy of life-enhancement, and it is our invitation to inner awareness.

When Thinking Arises from Knowing

As we feel attracted to the vibrations of our heart-consciousness, we can live in a state of knowing. This transforms our thinking processes. No longer do we need to think about survival or any negativity, freeing our mental processes to work on new and imaginative ideas that arise from our intuition. We no longer need idle thoughts. Absorbing ourselves in knowing our presence of awareness, we can allow our awareness to open into the timeless zone. This is the core space of creative consciousness, expressing itself through our personal vibrations.

On the way to knowing and deep understanding, we encounter every negatively-born limiting belief about ourselves, which have held us in fear and doubt, resulting in enslavement of varying degrees. When we resolve our attachments to limitations, we can be open and receptive to our heart-consciousness. This level of awareness is beyond the mental processes of ego-consciousness, and we must be willing to go within to the place of knowing and understanding. There are many spiritual practices designed to help us do this. If we are open and aligned with our heart-consciousness, we are guided to everything we need to be present, in absolute confidence with unconditional love and joy in every moment. We can live in the radiance of infinite love and realization of our eternal essence.

The consciousness that provides our personal awareness is

infinite and constantly expanding, as we and others like us constantly create new experiences of all kinds through our recognition and realization. With our free will, our thoughts and feelings, and our ability to realize what is real for us, we modulate the energetic patterns that we focus on, creating our own experiences. We live within our self-created frequency bands. When we expand our attention to more positive, higher frequencies, we no longer encounter negative frequencies.

Our heart-consciousness is entirely life-enhancing, and that's what our lives become in every way when we align with it. This alignment comes with inner knowing, divine guidance and great wisdom. Thinking arises from realization and becomes an expression of gratitude. From a perspective of heart-consciousness, we can play compassionately with ego-consciousness, while remaining unattached to limitations.

Opening to Infinite Awareness

Because it is infinite, consciousness cannot be measured. Our awareness exists within consciousness and is as expansive as we allow. As we open our receptivity to greater truth about who we are, our lives begin to change. We can feel and realize our energetic presence. This is not describable; but it can be known beyond the ego-mind. We can feel and realize the energetic presence of others, including those without physical bodies. How open and receptive we are, determines how expanded can be our awareness and alignment with our heart-consciousness. Our awareness can be beyond duality, without even a tinge of fear. This is in the realm of love and compassion.

By holding personal dramas and negative feelings toward others and ourselves, we block our awareness of the unconditional love in creator consciousness. By choosing to align with gratitude and joy, we can realize that we have an essence that we can feel. It is our presence of awareness. In the awareness of this presence, we can observe and direct our ego-consciousness, which cannot access infinite awareness and needs higher guidance to live a fulfilling life. We have been trained and have become accustomed to believing that we are our embodied ego-consciousness, and we have allowed this in order to have the most intense human experience.

No one can require us to live under limitations. They are entirely voluntary, but they will persist within us until we recognize and accept them and realize what they are. When we resolve and release them, we can open our awareness to our intuitive knowing. For eons we have lived with the belief that our limitations are real for us and for all humans. By carefully examining them and tracing them to their foundation in fear, we can find that that all limitation can be traced to belief in mortality. This is what we can transcend, when we realize that our presence of awareness is eternally present in universal consciousness.

In expressing ourselves as our current person, our human consciousness includes our subconscious, beyond the awareness of ego-consciousness. It is the storehouse of our experiences and limitations, the director of all of our bodily functions and the limiter of our awareness. Throughout our life, we have trained our subconscious to hold us in fear and limitation. We've also inherited limiting beliefs about ourselves. All of this can be resolved when we choose to open ourselves and receive the vibrations of Self-Realization flowing through our heart and enveloping us in the unlimited joy and ecstasy of life-enhancement for all.

Having the desire to know our true Self is the beginning of awakening. Then we must begin the quest of penetrating our limitations and opening our awareness to the infinite creative essence of who we are.

Living an Inspired Life

We all know intuitively that we are here to expand love within ourselves and among one another, but because we are coming out of an era of fear and oppression for our species, we are greatly challenged to actualize the enhancement of all life. Deep down we know we want to feel joy and love in every moment, and these feelings actually result from conscious choices on our part. We experience what our personal vibrations resonate with, and much of our lives transpire from the vibratory level of our subconscious essence without thought. Our conscious minds have not been fully present most of the time, and our bodies go through our daily lives according to our inherited and acquired programming that expresses our state of being.

When our conscious minds are in intentionally creative mode, our subconscious pays attention and adjusts its programming to our intentional vibrations. The more we do this, the deeper our intentional vibrations penetrate our subconscious. When we align ourselves in resonance with the vibrations of our heart center, we reprogram our energy signature to the spectrum of joy and love. With intentional energetic work, we can enhance this process. Our subconscious is not trained by thoughts, but by the qualities of our mental and emotional vibrations.

The polarity and vibratory level of the energy we express determines the quality of our experiences, in conjunction with the vibrations of our subconscious. When the subconscious polarity and vibratory patterns are dissonant from our conscious vibrations, our experiences become difficult for us. Experiences don't just happen for us; we create them with our magnetic radiance, which can now be measured by instruments that can register our magnetic vibrations at a distance from our bodies. These vibrations are emitted by our emotions, and they are what attract our experiences.

Without an out-of-body experience to retrain our subconscious, we must engage it as deeply as possible. In order to achieve our desires and goals, we must bring our subconscious into alignment with our desired quality of life, and then everything becomes easy. As we go deeper into our intuition to realize our true essence of awareness, we can vibrate at such a high level that we transcend our limiting beliefs about ourselves, and our awareness opens to a higher dimension within infinite consciousness.

Realizing Our Creative Essence

We can learn to lead with our heart and follow with our mind. Our heart-consciousness will provide us with what we want, but limited by what we subconsciously believe about ourselves. Once we can release our limitations, we are free to follow our heart and manifest the experiences we truly want in our lives. Our essence is an expression of the consciousness of the Creator of all, the One consciousness, which we participate in. We are our consciousness without limits. When we limit our awareness to the empirical world, we do not know our expansive Self.

When we intentionally open our awareness into alignment with the greatest love and joy we can imagine, we can open ourselves to our heart-consciousness. This is the center of our conscious life force and is the true power in our lives. It is our inner divine guidance and our potential genius in every aspect of life. It is our inner knowing without thought. Being able to transcend thoughts into a realm of mental and emotional silence enables us to align completely with our heart-consciousness. This state of being carries our ego-consciousness into a trance of observing and learning without stress.

In this clear state of being, we can examine our limiting beliefs and find their origin in hidden pain and suffering, while they keep us from being transcendent. By maintaining our awareness of heart-consciousness in compassion and love, we can accept our subjection to suffering. We no longer need that limitation, or any other based on belief in fear and mortality. We can be grateful for all of our experiences in deepening our understanding of negative energy and its effects, and we can release it from our inner knowing of ourselves as multidimensional, transcendent Beings with infinite awareness beyond time and space.

It's a big leap in expanding our conscious awareness to become heart-conscious in the enhancement of all life. When we do so, our awareness opens to our creative power, which we constantly use with our imagination and emotions. This ability can be greatly enhanced by clarity in aligning with our intuition. When we know our eternal essence, we can have absolute confidence in our creations. As Self-Realized fractals of creator consciousness, we have no limits on what we can transmit into the quantum field of all potentialities for manifestation in our lives.

Recognizing and Transcending Our Limitations

Because we have created, intentionally or subconsciously, every personal experience, we can learn to recognize how we do this, and how to improve it. Our willingness to doubt our inherent abilities has limited us to living in duality. As long as we believe in doubt, we introduce dissonant energies into our creations, keeping them from manifesting the way we want.

There are things we can do to replace our doubts with fulfillment. We can expand our awareness in increments by imagining what we want and choosing something better than where we currently are. We can start with things that we would not completely believe are possible for us, but that we can stretch our belief into realizing as real for us. Every kind of fulfillment begins in life-enhancing love and joy. When we can imagine and feel ourselves living more wonderful and satisfying lives, we can pay attention to the intuition of our heart-consciousness and feel ourselves living in alignment with the fulfillment of greater love.

This experience is beyond ego-consciousness and opens us to greater awareness of who we are within infinite consciousness. In this state of Being, anything is possible, because we have no limits. We are infinitely powerful creators in every dimension, but we cannot interfere with the consciousness of another self.

Although we've been living under difficult limitations for eons, we are not required to do so. It is voluntary. We are held within our limitations solely by our own beliefs about ourselves. These are very deep and hidden in our subconscious, and they are shared by all of us. They are based in fear and belief in our mortality. In order to resolve them all, we can develop acute sensitivity to our intuitive knowing and feeling. We have a conscious connection with infinite awareness, and to access it, we need to realize our connection.

Belief in our finiteness and inferiority has made us vulnerable to control. But it is only our belief that makes us vulnerable. Without our acceptance and alignment with limitation, we could not experience its energies, even if it is pervasive around us. Our powers of energetic choice and focus of attention cannot be impinged upon and are the initiators of our creativity in every moment.

Experiencing the fulfillment of our desires is a conscious choice. We can believe that we are limited in what we can manifest, or we can believe that we are unlimited. There is no difference between these beliefs, except for the results of what we experience. If we want to live in the energy of our heart-consciousness, we can begin by believing it is possible and then sensitizing ourselves to our inner guidance, which is our connection with greater awareness of creative joy and gratitude.

Opening to Our Transcendence

When we focus on our transcendence, and we hold our attention on just being present in awareness, our brain waves change into resonance with our heart-consciousness. We can feel the Source of our being in the infinite love and compassion of our heart, and we can open our awareness to a higher realm of energetic expression. It is a realm of greater passion and vitality. Here everyone is supportive of everyone in all of the diversity that we choose for self-expression.

As we realize our transcendent self-determination, we may become more consciously-directed in our thoughts and feelings, knowing that they are constantly creating the quality of our lives. Imbuing our conscious life-stream with our personal limitations, our subconscious pays attention to our vibratory expressions and receives the full feeling of our state of being. With practice, our present attention transforms our old subconscious programs of self-limitation.

When we are in the state of just being present in awareness, we can understand our situation without personal desire or need. Our reception and acceptance of love and compassion enhance our energetic signature, resulting in more love and joy in our experiences. With fewer limiting beliefs about ourselves, we can expand our awareness within universal consciousness. As our conscious awareness expands, we can align more closely with our intuition for continuous guidance. We can have an inner conversation with ourselves, learning to be aware of our heart-consciousness and knowing its feelings, which are immediate in every moment.

Following our inner guidance makes life easy and eliminates inner conflict and unnecessary drama, allowing us to be more intentionally creative with confidence in our abilities. Being present in a state of gratitude, we can know greater joy in all aspects of life. We can learn to confront every encounter creatively, rather than reactively. Every eventuality is always present, and we can choose any quality that we desire.

Being present in transcendence, we do not experience negativity. Even though it may be around us, it is in the realm of duality, a different energetic bandwidth from heart-consciousness. In a state of personal transcendence, all thought arises from intuitive knowing. Everything becomes an expression of life-enhancement, and we can realize our unlimited creative ability and our expanding presence of awareness. We can change dimensions in our awareness, and eventually we can take the essence of our physical bodies with us in the realization of our essence beyond time/space. This is an individual path for us on the way to realizing our infinite Self.

Anticipating Life with Confidence

As our traditional social, monetary, political and military systems are beginning to collapse, our more enhanced way of being is arising. We are being urged by nature and our greater environment to listen to our heart. There are crises all around, causing our ego-consciousness to be in fear of what's happening, because there is no apparent escape from utter chaos or complete enslavement. But we have other choices. If we choose to let go of fear and strife, we can give our attention to what we love and enjoy in the deepest ways.

Within our intuition we know and understand our current situation. We are held in fear and stress by our own belief system, based in fear of suffering and termination. We have believed in our limitations, but we can change our perspective by intentionally recognizing a different reality and realizing ourselves living in it. By recognizing and aligning with heart-consciousness, we can live well in any dimension. Everything is, in its essence, part of Creator consciousness.

All future experiences are created by our energetic resonance in how we feel about ourselves in every moment. Once we realize that we are infinitely powerful creators, and we are constantly using this power with our mental and emotional processes, we can be complete in ourselves. Our polarity and vibratory level determine our experiences. When we release fear and replace it with love, we change our polarity and the trajectory of our lives. We can feel that unconditional love is the quality of our conscious life-force.

Our essence is our presence of awareness, which we control with our attention. We create the negativity and positivity in our experiences by how we feel in every moment. It is our vibrations. This is what our subconscious is aware of and broadcasts through our aura into the quantum field. When we react with negativity or positivity, that is what we are creating in our future. We are energy modulators and creators of experiences through our imagination and emotions. This is the power we have been unaware of. We all have it, and we all use it all the time.

By becoming aware of our creative ability, we can learn to control our minds and emotions, so that we constantly express the desires of life-enhancement. We can live gratefully and joyfully in the eternal present, with awareness of everything, everywhere, all at once. In this state of Being, we can direct our paths in life to the most personally fulfilling ones. In every moment we have infinite choices for being ourselves and forming experiences. All opportunities are always present, awaiting our recognition to become real for us. We can reach out with our imagination and feelings, always anticipating greater enhancement of all life. We choose our own future, and we choose how we feel in every moment.

Opening To Expanded Awareness

An enlightened being is observant of the kind of awareness that creates our expressions. Paying attention triggers our creative imagination and emotions. Thoughts, words and deeds all have vibratory patterns in our consciousness. They stimulate us and receive our creative recognition. With all of these things being expressions of our conscious and subconscious minds, we can train our subconscious to align with us in awareness of heart-consciousness. In alignment with our entire conscious Self, we can be intentionally and completely present in our creative awareness. Our attention in this state can create personal transcendence, while we release the belief in our conscious limitations, opening our subconscious to our awareness.

When we are aware of our limiting beliefs and can identify them, we can resolve them by aligning our imagination and feelings with heart-consciousness. This energy is the source of our vitality and essence. It is always present awaiting our awareness.

Being able to observe our mental and emotional processes is a way of learning how we create our lives. We can be aware of how positive or negative we are, and we can choose to direct our imagination and emotions toward the life-enhancing energy of heart-consciousness. We know how energies feel, and we know how to align with the ones we want to experience.

It is possible to change every aspect of our lives, and we are being invited to do this in alignment with our heart-consciousness. There are many qualifying decisions that we must make.

We can learn to trust ourselves implicitly as we release negativity from our vibratory spectrum. We can become intentionally supportive of the renewal and enhancement of all life. In this state of Being, we can become aware of the expansiveness of our intuitive knowing. Once we realize the truth of our inner knowing, we have constant higher guidance to fulfill the desires of our heart in gratitude and joy.

We can be aware of all the chaos and fear in the human experience and give it no engagement, only compassion and love for all who are still entranced in the realm of duality. In following the guidance that we realize within the heart of our Being, we can live with all of our personal needs and desires fulfilled, while we provide creative energetic support for all. As our awareness opens beyond ego-consciousness, the realm of Creator Consciousness opens to us in its fullness.

Expanding our Realizations

Being able to focus our attention when we're quieting our mind and slowing our brain waves is possible by being present in awareness without distraction. It is a state of quiet ecstasy and unlimited awareness in alignment with the vibrations of our heart-consciousness. We can practice always being joyful and grateful, transcending our belief in vulnerability. Our awareness is eternally present. It is not bound by time and is the source of our creative intent.

We are always aware, even if our body is in a coma. We always have thoughts and feelings that serve many purposes. They are both receptive and radiant. In their radiance, our mental and emotional processes attract resonating energy patterns. When we meet someone in person, we know immediately if our vibrations are compatible. We feel their qualities, and we transmit ours.

In being receptive, our imagination and emotions connect with the vibrations we're facing. When we're in a slow brain-wave state, we are not being intentionally creative, we are just being present with high-vibrations. This allows our essence to express our deepest creative intent through our state of Being in joy, gratitude and love without having to do anything beyond our personal vibratory level. This is our best form of creative expression and results in our most fulfilling experiences.

The energies of this space of joy between waking and sleeping can stay with us and inspire us throughout our day, guiding us to align with the love and compassion of our intuitive heart-consciousness. Within this state of being, we are receptive to super-learning, and it is a powerful way to reprogram our limiting beliefs about ourselves, allowing us to open ourselves to our infinite conscious awareness and realize the potential of our creative ability.

We are such creators that we have created prisons for ourselves as well as palaces. We can expand our creative ability in conjunction with our ability to believe in what's possible. With practice we can expand our limits, until finally we realize infinity. Beyond everything, we can realize pure consciousness forever in the present moment. Every possible scenario and energetic pattern is available for our creative intent, all culminating in our experiences in any dimension we choose and with whatever quality we feel. We can have awareness of our infinite presence, while also being in space/time awareness. This gives us a perspective with understanding beyond human realization.

Transcending Our Empirical Trance

Being in the dualistic empirical trance of humanity is powerfully addictive; however, it is possible to expand beyond body-consciousness. In this state we can transcend our limitations and enter a state of greater awareness beyond polarity. In our emotional expressions we are naturally magnetic, attracting resonating patterns of energy. This is how we create the qualities of our experiences.

Expanding beyond body consciousness is a process of allowance, but just being present in awareness and allowing ourselves to expand beyond our physical presence receives resistance from our limiting beliefs about ourselves. We have believed that our reality is limited to the material world, and that we cannot go beyond empirical reality. We have believed that this is the only real dimension, resulting in blinding ourselves to others. This belief can be resolved by an out-of-body experience. Many have had this kind of experience and have reported about it in videos and books, but nothing tops personal experience.

It is possible to gain the same insights that result from death of the body through processes of deep meditation, ayahuasca, conscious dreaming and others. In conscious dreaming we maintain conscious awareness in the space that we cross between waking and sleeping. With practice we can open our awareness as we remain in the vibratory state of conscious dreaming. Here we can direct our visions and feelings in ways that our subconscious pays close attention to. We can change our deeply-set limiting beliefs in resonance with our vibratory alignment.

Entering an out-of-body experience can occur by putting our body and brain to sleep, while maintaining conscious awareness, which we can direct to the heart of our Being. We can be aware of the feelings that come to us through our heart, expanding into joyous, life-enhancing feelings that convey our conscious life force. In this parallel energetic dimension, we can experience whatever quality we choose with anyone we resonate with.

We can meet our angels and guides and feel the presence of our essence of infinite Being. When we open our awareness to the reality of a higher dimension of living, we are no longer enclosed in the limitations of the material world. We can direct the qualities of our experiences and even the forms, if we want. Once we learn to control our energetic alignment, our experiences may even appear miraculous.

"Be still and know that I AM God"

This state of Being is what comes to us when we release everything we think we know about ourselves, requiring complete openness in trusting our eternal presence of awareness beyond any beliefs that we are limited in any way. We are Creator Consciousness expressing ourselves as humans with inner awareness of the Source of our conscious life stream. We can call it intuition. It is the realization of our infinite essence of life. The essence of Creator Consciousness that we share is a fractal of the entirety of universal consciousness. Through our mental and emotional abilities, we are the creators of the expanding cosmos, by sending our personal radiance into the quantum field within universal consciousness. We are all the consciousness of the Creator of all.

It is possible for us to realize our expanded personal reality. We are expressing ourselves through our bodies for our learning with empirical experiences and negative energy. These were unknown for us prior to our first incarnation in this spectrum of energetics. When we learn to calm the mind and emotions into a state of neutral observation, we can be aware of our inner guidance, which is not intrusive. To be aware of it, we must seek it within ourselves by directing our attention to the most love and joy that we can imagine.

In order to open ourselves to our true intuitive knowing, we must align with the energy of the heart of our Being, the Source of our awareness. We have absolute control of our attention and the freedom to use it however we desire. We are learning to trust our Self in an environment in which we can't do too much damage, because we thoroughly doubt our abilities, disabling our creative intentions. We wanted experiences in negativity in order to enhance realization of our inner light and infinite presence of awareness. We knew that we'd become entranced in this limiting world, and we gave ourselves a way back to Self-Realization through our intuition.

Awareness of our intuitive guidance requires more than intentional searching for it. It requires recognition of its energetic expressions everywhere in our awareness. Paying attention to the feelings that arise in us in every moment, can convey to us realizations, knowing and understanding. With this guidance we can walk through the valley of the shadow of death and be unfazed, while we maintain inner awareness of our infinite Source. This is the ability that we are all learning.

Realizing Infinite Love and Joy

The consciousness that we all share is infinite, but the portion that we are aware of is localized within our chosen, accepted and realized limitations. In order to have greater awareness, we must recognize our limitations for what they are and realize that they are all based on beliefs that we have created for ourselves in alignment with belief in our mortality. We have believed in the inevitability of deprivation, suffering, aging and death. Through the vibrations of these beliefs, we have attracted experiences that resonate with them.

If we want to be more light-filled in our Self-Realization, we must resolve our limiting beliefs about ourselves. We created these beliefs for the purpose of developing greater compassion and deeper understanding of the effects of encountering and engaging with negative energy. Aligning with negativity is always life-diminishing. We are learning to identify negative energy and transform it to the vibrations of life-enhancement. We do this by using our mental and emotional powers in inspiring and loving ways in all of our encounters and experiences. This is not weakness. It is the most powerful way of being, because it establishes a personal vibratory quality beyond the experience of negativity.

There is no need to judge anyone as bad or inadequate. These are traits within our own consciousness, and we project them upon others, not having been able to accept them for ourselves. We have lived without higher guidance, and in many ways we have participated in things that we do not want to acknowledge. We must choose to forgive ourselves for our inability to love ourselves and everyone unconditionally always. Only when we are perfectly in alignment with the energy of our heart can we have this ability, because we can transcend our limitations by realizing our eternal presence of awareness and infinite creative power that constantly arises for us within the life-essence of universal consciousness.

Once we realize that our physical presence is an expression of our essence, we are free to change everything about ourselves. All we need to do is expand our realization of what is real. Within our own awareness we have the ability to feel, know and realize our expansive Self. It requires our intentional awareness in gratitude of the great joy and love constantly arising and radiating from our heart-consciousness. This is the transformative energy that lives within us and guides us to experiences that we deeply desire. Expanded realization and knowing is a gift that comes from opening to heart-consciousness.

The Origin and Resolution of Personal Limitation

From within the limitations of our awareness, we cannot know our true essence. We are beyond time/space in eternally present infinite awareness, expressing ourselves as humans living in a compartmentalized spectrum within universal consciousness. From within the consciousness of humanity, we do not know unconditional love and abundance. If we want to know this kind of life, we can begin by filling ourselves with gratitude for every experience.

We can have the perspective that we have created our current situation for our learning and fulfillment of something that we feel encumbered by, or for something that we love. When we have the feeling that we are enhancing ourselves and all conscious beings, we are resolving our ego-conscious fear of deprivation. We are allowing the expansion of personal awareness into a greater understanding of the cause of our experiences.

All of our personal limitations depend upon beliefs that we hold, and that we can intentionally change at any time. These beliefs arise from some form of fear that we do not allow ourselves to realize. We can trace this fear to concerns for survival and comfort. These concerns are held in ego-consciousness and develop into fear of the unknown. Unless we resolve this fear, we cannot expand our awareness. This is where we can work with ourselves for an inner awareness of heart-consciousness.

Our feelings always inform us of the presence of negativity or positivity. When we face negativity, it is because we have engaged with it in the past, and we now have the opportunity to transform our vibratory alignment. Our experiences are reflections of energetic patterns that we hold in our consciousness, including our subconscious. By shifting our vibratory alignment to the energy of our heart-consciousness, we transform our outer experiences.

Nothing is actually "outer." Everything is contained in our consciousness, and we are capable of infinite awareness. Everything that we experience is patterns of energy that our consciousness interprets as empirical stimulation of our senses. It is all going on within universal consciousness, which we participate in. There is only one consciousness, and we are aware of as much of it as we allow ourselves. Held within human awareness, we have been our only limitations, and we are being invited to resolve them and free ourselves to live in a state of gratitude, joy and fulfillment for all.

Conscious Light Is Enveloping Our Awareness

The reality we recognize as humans is constantly stimulating our physical senses within the energetic band of polarities and frequencies of the dualistic world of empiricism. We are limited to this version of reality by self-imposed beliefs about ourselves. There is no reality for us beyond our beliefs, except for the intuition of our heart. This is where we are connected to consciousness beyond the ego-mind. It is the consciousness of constant creation of new life and experiences, shared by all conscious beings. To access our heart-consciousness, we must intentionally align with its feelings and sense of knowing by calming our thoughts and emotions and allowing the expressions of our heart to pervade us.

All conscious beings emit an electromagnetic radiance, including photons of light. If we allow ourselves to feel this radiance, it is present for us. Every ability that is hidden in our subconscious can be available to us, if we can realize it intuitively. Our true essence expresses itself through our heart-consciousness.

What we perceive as outside of us is occurring in our own consciousness, and is being presented to us as a review of what we imagined, feared and loved. The cause of our experiences is the patterns of energy that we vibrate at. How we react to experiences determines our vibratory expression and provides the creative intent for our new experiences. If we can stay aligned with our heart-consciousness, all of our abilities begin to open to us.

Regardless of what may appear to be happening outside of us, our personal expression of heart-consciousness is what enables us to create life-enhancing experiences. Our vibrations are our creative expressions that interact with any energetic patterns that we pay attention to. As long as we're aligning with duality, our interactions are positive or negative, and they return to us as the qualities of our experiences. In the realm of duality, the expressions of heart-consciousness are always positive. When we align with our deepest gratitude, love and joy, we are in the presence of our heart-consciousness. This is our conscious life-force flowing to us in creative love within infinite consciousness.

As we become aware of our heart-consciousness, and pay attention to its stimulations, we open ourselves to greater awareness of our infinite presence and unlimited life-enhancing abilities. We can be aware of the light-being in everyone and everything that exists. It is all our own Being and our connection with universal consciousness and the Being it arises from, and whom we are fractals of in eternal presence of awareness.

Examining the World Within Our Consciousness

Opening ourselves to the subtle prompting of our heart-consciousness is the portal to expanded conscious awareness. This is the source of our essence arising within universal consciousness. It carries the vibratory expressions of Creator Consciousness, endowing us with conscious life-force and filling us with its vitality and infinite love. Without awareness of heart-consciousness, we cannot be aware of our source energy and our essential Being. We need to know what it feels like.

We all have intuition arising within our heart, and connecting us to the fullness of infinite consciousness. When we are able to align with these energetics, their qualities arise in us without limitation, and we become intentional creators of our entire lives, always knowing the most life-enhancing path in all interactions and visionary experiences. In this state of Being, we are completely free and sovereign, confident in creating fulfilling experiences of everything that is important and passionate for us.

Knowing intuitively requires a quiet mind and a calm emotional state, so that we are not distracted from our inner prompting in every moment. Our inner guidance isn't usually verbal. It is symbolic and emotional in some way, and it is what we always deeply know in the immediate present moment. Once we have transcended our limiting beliefs about ourselves, our attention can go to scenarios that feel wonderful and life-enhancing. In this state of being, everyone is living in gratitude, joy and abundance without limit. This is how our creativity blossoms into manifesting our greatest love.

We are constantly creating the qualities of our experiences by our mental and emotional vibratory states. To the extent that we dwell in limitations, we suffer, grow feeble and die, at least in the dualistic, empirical world. Once we are free of physicality, we can more easily remember who we are as Source Beings. If we want out-of-body experiences, we can have them show us the truth of our essence, but we don't need these experiences to become heart-conscious.

As we learn to be in gratitude and greater awareness, we gain transcendence and access to the infinite love and joy of universal consciousness. In our creative ability, we become able to change the vibrations within ourselves and in our energetic environment with our imagination and emotions in confidence and joy. We enter the realm of cause. Through our limitless love, we naturally manifest our heart's desires in all of life and in every experience. This is the world we can create for ourselves and all of humanity now.

Becoming Absolutely Free and Sovereign

By living in a state of transcendence in awareness beyond good and evil, we can become masters of the human experience. We can live in our eternal presence of awareness with unlimited creative power for the enhancement of all life. This is the state of Being that we are being attracted to by the enveloping consciousness all around us. We are participants in the consciousness that constantly creates everything trillions of times per second.

In realizing that we are fractals of universal consciousness, we can know that we have always been the creators of our experiences in collaboration with all other humans and everything we interact with. We have our species consciousness, just as the animals and plants do. The Earth has her planetary consciousness, which is becoming more powerful in recovering her creative life force, as she rises in vibratory resonance. To continue to live here, we must align our state of being with her. We can benefit by being in nature as much as possible, often by ourselves in places of beauty and majesty, with our skin on the earth when possible.

In order to be able to utilize our unlimited creative ability, we must live in harmony with our environment and everyone around us. We must align with the qualities inherent in universal consciousness. These are the qualities of infinite love and joy with complete confidence in our abilities. It is primarily fear and doubt that have restricted our awareness and realization. When we know and feel in alignment with life-enhancement for all in every moment, we can trust ourselves implicitly, and our abilities become unlimited. We can resolve and release our limiting beliefs about ourselves.

In realizing our presence of awareness beyond time/space, we cannot be threatened or invaded by negativity. It does not exist for us, because we are in different energetic dimensions. In our human lives, this results in ultimate freedom and personal sovereignty, regardless of what may be happening around us. It is all within our consciousness, and we can choose which vibratory patterns we want to pay attention to and align with. This is how we create our quality of life, and we can experience what we love the most.

Adapting to Our Changing Cosmic Environment

Although we exist as individual expressions of infinite consciousness, we appear as our own creative expressions in multiple dimensions. Each of us is in essence a personalized presence of awareness with infinite creative power that we are designed to use as we choose. Since we are participating in a dualistic world that we believe is real, fear and doubt keep us from trusting ourselves with our abilities. We have accepted beliefs that limit our expressions. This keeps us from destroying the cosmos while we attempt to survive without higher guidance.

Because of the rising positive energies of our cosmic environment in the form of conscious gamma rays penetrating everything with fractalizations of universal consciousness, we are being drawn toward raising our vibrations into expressions of gratitude and compassion. The dimension containing negative energy is dissolving, and we are entering a dimension beyond polarity, in which all energetic patterns are life-enhancing. This is the quality of energetic expressions throughout the larger portion of universal consciousness, and is the expression that comes to us through the intuition of the heart of our Being. We have as much access to it as we allow ourselves.

Through our imagination, we create our own reality. We believe that the energetics of the empirical world are solid and physically experienced. In essence they are patterns of electromagnetic waves that our consciousness interprets through our imagination as our empirical reality. All of these waves are comprised of minute conscious beings, and we can feel their vibratory expressions. By focusing our attention on certain energetics, we bring them into our recognition, and we interact energetically with them in a way that we realize as reality. Without our recognition, they remain part of our energetic environment, but they are not in our awareness.

We can expand our awareness in any way that we desire. Once we have been able to calm our emotions and mental processes to a state of clarity, we can have complete control, consciously and subconsciously, of our imagination and emotions. We can be truly creative in expressing our passionate attractions in life-enhancing ways. We are the creators of experiences, and they can be as wonderful as we imagine with gratitude and joy. As we pay attention to our visions and feel ourselves living in them, we change our personal energy signature in alignment with their qualities, attracting experiences that resonate positively with us. Eventually we can realize that we are experiencing the quality of the realm that we create in our imagination.

Being the Persons We Love the Most

We have been programmed to believe that we are separate beings with a separate consciousness, and this is how we perceive ourselves to be. Because we have believed that we are our bodies, we have also believed that there is a world outside of us. The world we believe is outside of us is filled with drama of all kinds, good and bad, but ultimately our participation becomes lethal for the body.

The expansion of our scientific abilities through technology and quantum physics has given us a deeper understanding of our condition. Our consciousness is not localized to us. Only our awareness is. Consciousness is infinite, and we are expressions of it. We constantly arise within it. It is our essence of being and our conscious life force. We cannot be separated from infinite consciousness, but we can block our awareness of it with our limiting beliefs about ourselves.

The entire play of the empirical world is constantly being created within our own consciousness as we interact with other conscious beings. Everything is conscious, or it could not exist. This includes subatomic protons, quarks, neutrons, electrons and photons, all of which display consciousness and comprise our bodies and our entire experience. Everything in our experience is conscious, which means that it all exists in universal consciousness. All of this has been discovered and logically deduced from empirical experiments that anyone can duplicate and get the same results.

With our limiting beliefs, we have been entranced in deluding ourselves for the sake of having experiences in limitation within a realm subject to negative experiences. When we are ready, we can open our awareness to our true Being and realize that the entire empirical show is happening in our consciousness, and we are its directors, producers and actors.

Everything is electromagnetic energy, which we can feel, envision and modulate. We create patterns of energy with our mind and emotions, and we align with vibratory qualities through our attention and feelings. We don't have a choice about this. It's how we are created. We are fractals of Creator-Consciousness with the freedom to use our minds and emotions however we want. The quality of everything in our experience is a vibratory expression of how we feel about ourselves and who we believe we are.

Although we have not been aware of it, each of us encompasses the entire cosmos, free to create whatever expressions we desire. Nothing can force us to suffer or worry. These are personal decisions, and they create reflections of themselves. As the source of our conscious life force, our heart-consciousness is the logical and emotional place to pay attention to. It is our connection to infinite consciousness and our essence of Being. We are the creators of our life expressions and everything we experience. Through the vibratory guidance of our heart, we can live in infinite love and joy.

Rising into Our Magnificence

As humans, we have been blind to our true essence. We share the belief that we are our ego-consciousness and our body. From within the limitations of ego-consciousness, which we have imposed upon ourselves, we cannot expand our awareness to the greater consciousness that we participate in. This greater consciousness is inviting us into awareness of it. In our essence we are fractals of infinite consciousness. We encompass the cosmos within our true Being.

Infinite consciousness conveys feelings and stimulations beyond our current experience. When we open our awareness to them, we can realize that our human reality is an expression of a much greater Being of infinite consciousness. Out of this consciousness we arise as personal expressions. Our Being is the Source of all conscious life everywhere, and we are the creators of everything we experience.

For eons humans have lived within the limitations of empirical duality, not realizing that this is all imagined within our consciousness. Without our conscious life force, which we convey through our energetic alignment in our awareness, it could not exist. The reality of the world that we have experienced depends upon our recognition and energetic alignment. It all takes place in our imagination, and we can change the quality of it with our intentions. We are imagining the world of human experience. It consists of patterns of electromagnetic waves that we perceive in our subconscious awareness as the empirical world. We have inherited this and learned it from one another.

Because we have believed that we can be intimidated and terminated, we could not believe in infinite awareness. If we desire to realize our greater Self, we are guided from within, transcending our limiting beliefs. We can realize that every thought and feeling is important, because it reflects back to us in our experience. We have made ourselves as we are, and we can transform our current state of being in alignment with our higher guidance.

Our inner guidance is what we know and feel in our essence, which is universal consciousness. It has the vibratory quality of the enhancement of life, including the greatest love, joy and abundance. The human ego cannot imagine this as our reality, but the ego-consciousness can be transcended by gratitude with inner knowing and feeling. We are coming into realization of our eternal, infinite presence of awareness, with the ability to create whatever we desire.

Expanding into Greater Realization

As we go about our lives, we can be inspired to realize greater truth about ourselves. We have the ability to fill ourselves with gratitude and joy in every moment. By doing this we vibrate at a frequency that transcends our self-imposed limiting beliefs about ourselves, and we enter a dimension of oneness with one another in compassion and understanding beyond ego-consciousness. All around us the energies of life-enhancement are increasingly supportive of our appreciation and alignment with our heart-consciousness.

We are being invited to expand our awareness to the deepest knowing and feeling arising in us within universal consciousness, as it provides the essence of our Being. By identifying with the grandest, most loving and joyful feelings, we come into resonance with our heart-consciousness and our inner knowing. Intuitively we come to know the feeling of infinite love and deep connection with everyone, as we become aware of our species consciousness. We all arise within universal consciousness as the same etheric fractals with free will to create ourselves however we desire, beginning prior to our incarnation, according to the kinds of experiences we plan to have and the qualities of our relationships. It all happens on an energetic level and comes into our awareness in the vibratory patterns of the empirical world.

We have formed limiting beliefs in order to enhance the

reality of our human experiences, which we could not seriously have, when we are aware of our infinite, eternal consciousness. We have the ability to know our true Selves and our limited human selves, in which we can become aware of our inner light, and the radiance of everyone in our field of awareness. Although our physical forms become less meaningful for us, our bodies do become more beautiful and attractive. We have an energetic effect on our personal consciousness that grows in power and influence without personally-limiting beliefs.

In just living our lives, as we become more heart-centered, our energetic focus strengthens, and our attention becomes more powerful, resulting in faster and more intense creations. Our entire lives transform into expressions of love and joy. This process begins with our intentional alignment with the life-enhancing energies of our heart. The more we can feel and identify with heart energy, the more expansive our realization becomes, until we can realize our essence as infinite presence of awareness beyond space/time and including the entire cosmos and potentially beyond. We are the creators of all of it within universal consciousness. Creation happens through us. We create experiences, and we are doing so effortlessly in every moment now and always. As we open our awareness to higher qualities of life, we can use our powers of energetic modulation intentionally to enhance all life, including ours.

Transforming the World of Our Experience

Within the space/time continuum, it is possible for us to choose to align with vibrations that give us more vitality and exuberance for life. To do this successfully, it can be helpful to be aware of the source of our conscious life force within ourselves. This we may identify as our heart-consciousness, expressing Creator consciousness through all of the energy centers in our body and etheric presence. This is our natural way of Being. Except for our intention and openness, we do not need effort to feel and live in our natural vibrations. These are natural for us beyond our time/space enclosure as well as within it. Through the natural vibrations of our heart-consciousness, we receive and radiate the energies of abundance, joy, freedom and love. When we allow ourselves the experience of these vibrations, we are aligning with our true essence of infinite consciousness.

By aligning with our Source vibrations, we no longer experience negativity, which stimulates our feelings as fear. When we no longer align with it, fear no longer exists for us, because it needs us to create it for ourselves. By following our intuition, we are no longer subject to conditions that stimulate fear, because they are in a different, parallel dimension that we do not need to engage with, even though we may be aware of it. Without emotionally interacting with it, we do not give it our life-force, making it unreal for ourselves. In order for us to experience this kind of alignment, we must release all doubt about our abilities and replace it with continuous confidence, gratitude, joy and love, our natural state of Being.

Making the transition from a life of ego-consciousness struggling to survive, to a life of gratitude, joy, love and abundance, requires transformation of awareness. It requires confidence and trust in our heart-consciousness and intuitive knowing, which comes to us as realization. Through our realization, we create our reality, subject to any limiting beliefs about ourselves. As we learn to trust ourselves to enhance the energy within and all around us in alignment with our heart-consciousness, we empower our creative visions and feelings. Once we are able to command every situation within ourselves, we can become energy transformers and shapers of our reality.

To transform negative energy patterns in the world of human experience, we can change the energy within ourselves. By seeing the light in any negative person or circumstance, we allow the negativity to become unstable and dissolve. We can modulate the energies in our thoughts and feelings into alignment with heart-consciousness. By transforming our own awareness and energetic alignment, we transform the world of our experience.

Observations on Realizing a Greater Reality

Within infinite consciousness, we arise as fractals of Creator Consciousness, unlimited in every way. Although we participate in infinite consciousness, we have limited our human selves with fear and doubt in order to participate in a realm that has been impossible for us to experience. Now we know what negativity feels like, and what it creates in our experience. We are living with the results of our mental and emotional vibrations. With our limited awareness, we are able to realize the reality of the world of human experience, and we've become so entranced in it, that we have been unaware of what is beyond.

When we become curious about the essence of our life, we usually start looking for a teacher. Unknowingly, each of us has our own inner connection with infinite consciousness, and this is what we can become aware of. As long as we believe that we can be victims and can suffer and die permanently, we cannot open ourselves to infinite consciousness. We cannot believe that it exists for us, even though quantum physics has shown that it does. Except for personal experience with negativity, we do not need limiting beliefs about ourselves. They are purely a matter of choice, and we can change our preferences.

In order for us to adhere to our preferences, we must focus our attention with intentionality. Our ego consciousness constantly runs scenarios through our minds and emotions, wanting things that we keep ourselves from believing are possible. In actuality, everything is possible for us.

We live in a cosmic plasma of electromagnetic wave patterns that convey conscious impressions, some of which we recognize, align with and interact with. These express the energetic qualities that are important for us, but there are many more beyond our recognition. How we feel about ourselves determines our vibratory alignment, and it influences what we pay attention to. By believing and feeling that we are complete creators in our own awareness, we can create what we want in a heart-felt way. Our experiences are a result of our inner state of being. They are reflections of our thoughts and feelings, our beliefs and our perspective.

If we penetrate deeply into our inner essence, we can begin to experience our pure presence of awareness in every moment. We can recognize our ego-consciousness and our heart-consciousness. We know how they feel and the kind of experiences that they create. Instead of living in the hypnotic trance of humanity, we can choose to expand our awareness in ways that interest us and that we enjoy. There is no judgment of how we use our power of freedom to think and feel however we choose. There are only results. Either we limit and enslave ourselves to negativity, or we free ourselves in gratitude from limitation and allow ourselves to be guided by the intuition of our heart in infinite consciousness.

Standing Powerfully in the Face of Fear

In the state of global humanity, there is much fear for survival and great attempts to resist the diminishment of our lives. The governments of every nation have become lethal forces for their citizens. People fear this intensifying development. As long as any of us is in alignment with this negative energy, we continue to experience it. But if we change our energetic alignment to gratitude and love, we are in a different energetic dimension, and fear is no longer part of our awareness. Instead, the more powerfully we are grateful for our awareness, we can turn our attention and alignment to what we love in every situation that we're involved in.

We experience our energetic creations in every moment, and we also create our continuum of experiences by our current mental and emotional states. By enjoying, accepting, resisting or identifying with any energetic patterns or scenarios, we draw our awareness into alignment with them. This is the quality that we are creating in the moment for our experience.

Scientifically, everything consists of patterns of electromagnetic waves that we perceive as physical in our awareness. The empirical world that we are all aware of, is a kind of hologram held in our consciousness by our telepathic alignment, and is interpreted by our recognition as our reality. All of this exists within our own expanded consciousness. Consciousness is our essence, and it exists and constantly creates everything everywhere. Consciousness is our prime source of being, and it is universal. We are aware of as much of it as we allow ourselves.

When we open our awareness to greater consciousness, focusing on feelings that arise by being clear and present, we can identify energetic patterns that are life-enhancing situations that feel wonderful. This is an energetic alignment with our intuition. When we are able to resonate consistently with gratitude, compassion and joy, we can resolve and release our limiting beliefs about ourselves. We become aware of our eternal presence of awareness, always filled with the guidance of our heart-consciousness in conjunction with infinite awareness.

Before the creation of humanity, our presence of awareness always existed, and we will continue to be our presence of awareness beyond the time of humanity on this planet. We can interpret our experiences as humans in any way that we want. Everything can be a blessing or a curse, depending upon our perspective and beliefs. We experience the results of the qualities of our thoughts and emotions, regardless of anything that appears to be outside of us.

Experiencing a State of Knowing

Our subconscious mind has a strong connection with our heart-consciousness. When we come into alignment with our heart-consciousness, we also do so with our subconscious. As our conscious servant in every moment, its treasure is ours to claim in alignment with our heart. Our subconscious directs our body consciousness to express the vibratory qualities of our predominant thoughts and feelings through the cells of our body. Our subconscious doesn't understand us, since it has only deductive reasoning, but it records our vibrations and radiates our personal energy signature within and all around us, attracting resonating vibratory patterns.

When we align with negativity or positivity, it is reflected back to us as a symbolic message that is either stressful or enjoyable, and in some way we experience it. It may be of a different form than we imagined, but it has the same feeling. Since we can direct our emotions and thoughts, we can direct the polarity and frequency of our own vibrations. We can begin by learning to just be present without focus, until called upon. In its anxiety, our ego-consciousness keeps streaming scenarios through our awareness. We can just let them pass without our interest and alignment. Eventually, when we begin to follow our inner guidance in the consciousness of our heart, we can have inner silence. This is when our subconscious pays attention to our direction and brings us into alignment with our deeper Self, while calming our ego-consciousness, which becomes an observer.

As we understand thought, our subconscious does not think. It is aware of our mental and emotional energetics and knows how we feel about ourselves always. We imprint our subconscious with our vibrations, expressing our creative energy through the cells of our body as well as in the electromagnetic plasma around us.

Not needing thought, our process of knowing happens in our awareness and realization. In the dualistic world of humanity, we have restricted ourselves to limited awareness and realization. There are many techniques that we can use in learning to be our present awareness. Free of internal chatter and drama, we can have clear perception and knowing. The feelings that come with inner knowing are gratitude and joy, expanding into great understanding. By aligning ourselves with these energies, we open our awareness to our inner knowing. Once we recognize what it is, it can be indelible in our awareness. We can become free to be our true, infinite Selves, while transforming our human lives into experiences of gratitude, joy, compassion, abundance, freedom and whatever we truly want for ourselves and all of humanity.

Aligning with Our Conscious Essence

Living in infinite consciousness, each of us is a participant in the whole. In the heart of our Being, we can have awareness without limits throughout universal consciousness. The bodies that we believe are real for us, are created in our consciousness as expressions of energies that can challenge and invite us to attune with our intuitive knowing. The world of human experience is a game or a play that we are acting in. It's like a computer game, and we direct our acting in the quest to free ourselves from limitations.

As we began acting in this play, we did not know that we are the directors of our lives, or that we create the qualities of our experiences. Our state of being is a result of how we feel about ourselves. This is a matter of free choice in directing our thoughts and feelings. In our energetic presence we interact with potential experiences that are magnetically attractive, and we choose our own vibratory resonance in every moment.

In the play of our human experiences, there is a vast number of possible scenarios. We have absolute freedom of choice to determine the path that we take in every moment. We can control and direct how we feel in every situation, filtered by our limiting beliefs about ourselves. By resolving and releasing our limitations, we become clear in our presence of awareness. We can release or intensify all personal drama, resulting in experiences that give us the same kind of feelings.

If we are aware of the qualities of our state of being in each moment, we can direct our visions and emotions in ways that feel best for us and for all life. If we choose this path, we naturally receive whatever we need in every moment, because we are participating in the enhancement of life, including our own. Without internal stress from doubt and fear, we can be aware of our deepest knowing and can align ourselves with it in our mental, emotional and physical processes and expressions.

As we align with our heart-consciousness, we can open our awareness to the freedom that we have as creators of our life experiences. All the props in our play are provided for us to be able to act out our parts in resonance with our energy signature, which we constantly create. We can be as limited or unlimited as we desire, depending upon the kind of experiences we want.

As humans, we have the potential of greater awareness through our intuitive knowing in the consciousness of our heart. This consciousness is unlimited. Once we realize our essence beyond time/space in infinite conscious awareness, we become the masters of our lives in every dimension that we choose to participate in.

Training Ourselves for Greater Awareness

When we choose to be carried in joy on the vibrations of Creator Consciousness, we enter a space of unconditional love. This is unknown in ego-consciousness, because there is no element of fear. Ego-consciousness is based in fear of suffering and termination. Intentionally choosing gratitude and joy as our state of being, enables us to become aware of transcendence beyond ego-consciousness. In this realm we can realize ecstasy and fulfillment in every way. This energetic field enhances all life. For us to be aware of this dimension of reality, we must realize it in our inner knowing. This realization happens when we open our awareness intentionally and without attachments to limitations.

Limiting beliefs about ourselves hold us in ego-consciousness. In Creator Consciousness there are no limitations, and our entry into realization of its reality is personally transforming. Our entire lives change for the better. We do not need to be attached to anything, because we can create everything we want in every moment. From fear we can switch our focus to love, and we no longer provide our life force and alignment to fear and threats. We are completely cared-for and are beyond negativity. We can realize its unreality in the face of our eternal presence of conscious awareness with unlimited creative ability.

Even without awareness of our creative ability, we are still creating the qualities of our experiences by our thought patterns and emotional attachments. This is our nature, and we can direct it intentionally. We can train our subconscious mind to align with the nature of our heart-consciousness and work creatively with us. When our entire consciousness is in alignment with our inner knowing, we concentrate our creative power, resulting in immediate experience of our creative intentions.

Each of us is unique, and we get to determine the quality of our thoughts and emotions in every moment. This is our creative energy, and it reflects back to us as the qualities of our experiences. It all depends upon how we feel about ourselves. The polarity and frequency of our thought patterns and feelings about ourselves and our conditions magnetically attract resonating energies in the quantum field. All possibilities always exist for us to experience through our recognition and feeling. Through our imagination and emotions, we can bring anything into our realization, when we align in clarity with our intuitive presence of awareness.

Living Beyond Our Limitations

We have the choice of living in vibrations of enhancement of all life, including our own. When we do this, negativity disappears from our lives. In order for the chaos and life-diminishing energies of the world to affect us, we must give them our attention and engagement. When we fill our awareness with joy and love, we are in a higher energetic dimension beyond negativity, and we do not have negative thoughts, feelings and experiences.

By our social training and the perspectives of others, we have been held back from realizing our ability to live fully. We have accepted a reality of limitation. To begin resolving our limiting beliefs, we must realize that we have them, and no one enforces them, except each of us. We do it to ourselves. By realizing what our beliefs do for us, we can decide if we want them. What happens when we realize that beyond our limiting beliefs about ourselves, we are our eternal presence of awareness? We express ourselves as our human persons, but in our essence of Being, we have unlimited creative power in any dimension that we want to experience.

By living intentionally in making things better for everyone in our awareness, we can recognize the light of our own essence in the essence of everyone we encounter. We all share the same consciousness and can be aware of each other's awareness. In order to live together in love, we must be true. When everyone is aware of everyone's vibrations, we can realize our unity in essence and the consciousness that we all share.

As a race of billions of individuals, we share a world of vibrations that are made real for us as physical experiences. We create this world for ourselves by recognizing its energetic patterns. Our recognition causes an instant interaction that we perceive as our reality. Now that a significant number of us have decided to follow our heart-consciousness, humanity is coming into alignment with this level of energetics.

Our galactic environment is moving beyond negativity, and we are interacting with this environment. The Earth is becoming more filled with vitality and is requiring humanity to participate in renewal of life on this planet and removal of toxins. Whenever we are ready, we can each choose to live in the consciousness of our heart and be guided by infinite consciousness in joy and love.

Realizing the Mystery of Life

All of our higher internal energy centers are designed to expand our conscious awareness into full Self-Realization. This process begins in the heart. Our heart-consciousness is the most powerful in our essence and in our bodies. By realizing what our heart-consciousness is, we can align with it, and in doing so, we open our awareness to our higher energy centers that expand our awareness beyond time/space. Their vibrations guide us, as we open ourselves to infinite consciousness. This is our home and is where we are completely fulfilled and cared-for. It is all within our consciousness, and we can access it. In this state of Being, we can express ourselves as our human persons, knowing everything we need to know to be able constantly to express ourselves in ways that enhance life all around us.

This is our new world. We are living in it, whenever we can align ourselves with our heart-consciousness and realize how it feels and what it conveys to us. At this positive vibratory quality, we are beyond negativity and do not experience it. We experience the energy that we resonate with in our mental and emotional processes. In our expanded awareness, we transcend our ego-consciousness. Since ego is based in fear, it does not exist in the realm of love and joy. In this energetic dimension, we are our expanded Selves, realizing the infinite creative love flowing into us in every moment and empowering us in every way that we desire.

In the world we have been living in, our awareness has been

severely confined by fear, anger and doubt. All of our institutions are based in these energies, which cannot be in resonance with the new world. This means that we must restructure everything in resonance with heart-consciousness. Our government must truly serve the good of all, and all of our social structures must enhance our lives, so that all of us can continue to expand our conscious awareness in gratitude, love and joy.

In order for the new world to become dominant for us, we need a significant number who already live in the realm of heart-consciousness. This is a powerful force that is dissolving the world we have lived in. This is why we have so much instability and chaos everywhere. The negative energies that have governed our societies are now being revealed for what they are, and we are ready to let them go into another dimension.

We are being invited to realize the new world of love and joy by aligning with our heart-consciousness. This is the energy of our deepest Being, beyond any limiting beliefs about ourselves. It is our inspiration and source of everything. It is so obvious that we've been blind to it. It is the energy that we live in and that gives us life. It is like air for terrestrial creatures and water for marine species. It is our complete environment within and all around beyond time/space. In our deepest Being we know what it is. We just need to open our awareness to it.

Transcending Ego-Consciousness

One way of transcending ego-consciousness and opening our ourselves to realization of greater consciousness, is to spend time in a quiet and majestic part of nature while spacing out and imagining scenarios of great love and fulfillment. As we become acquainted with these vibrations, and being able to feel them, we are expanding our awareness of a greater world of exploring our consciousness and realizing how we are designed to live.

We can recognize when we disallow ourselves the emotions of our heart. This happens every time we go into some form of fear, when we diminish our life force with stress and negative feelings. If we find ourselves in this condition, we have the opportunity of directing our attention and emotional sensitivity to gratitude for everything, compassion for ourselves and joy for being aware of this.

Regardless of our condition, we do have the ability to choose our state of being. We are the directors of our awareness. Any limiting attachments that keep us from realizing infinite awareness within universal consciousness can be intentionally released. Since all attachments are based in some kind of fear, we can transform that energy into gratitude and joy, regardless of how we believe negativity arose in us. These are intentional choices. We may need to learn procedures that help us direct our attention successfully to the energies we love, especially if we've fallen into addictions to negative patterns.

We naturally feel and know how to be and how to pay attention to positive vibrations, and to resonate with them. This opens us to the energy of the heart of our conscious essence, expressed by our physical heart, which constantly radiates unconditional love for us, regardless of what we do to it. When we can realize what these energies are, and can align with them, we can realize a greater reality in every present moment.

When we can be in gratitude and joy, we can live in awareness of our conscious unity with all living creatures. We encompass everything within our consciousness, and we can be aware of the awareness of anyone we focus on. When we align with our heart-consciousness, we naturally enhance the lives of everyone we encounter, and we feel the love, because it's all happening within our consciousness and our presence of awareness. Our entire reality is within our consciousness, and we direct its quality, and to some extent its form.

Recognizing, Resolving and Releasing Our Limiting Attachments

We are the masters of our beliefs. We have accepted and created them to define our reality. They limit our awareness, so that we cannot know who we really are in our essence. Beyond the negativity and fear of ego-consciousness, we can be free and sovereign and able to live in love, joy and abundance. For this we must realize our infinite conscious essence in the heart of our Being, our eternal presence of awareness. This is where we can realize that everything we have regarded as empirically real is an energetic expression of consciousness, arising within the essence of infinite Creator Consciousness, of which we are fractals, with infinite creative power and ability. This power has been denied to us, because we could not trust ourselves apart from the energy of our heart.

By observing how we feel about ourselves in any moment, we can be aware of any limitation that keeps us from realizing our infinite essence. Upon recognizing a limiting belief, we can trace it to its source, find the fear that it is based on and resolve it in forgiveness and gratitude for the experience granted to us. We are thoroughly schooled in what all frequencies of negative vibrations feel like.

Doubt and fear are the opposite of creative energy. They are diminishing and destructive to life, making us feel small and insignificant. As long as we hold onto them and their resonating energy patterns, we allow them to limit us to beliefs that are not true about us. They only define the roles that we are playing. Without limiting beliefs, we have no ego-consciousness. We have only our infinitely powerful and creative Self-Realization of Being.

From unlimited awareness, we can realize the essence of the empirical world and its purpose of giving us intense experiences of energetic patterns that we had not experienced in unlimited awareness, because negativity can exist only in a limited dimension. Within this dimension it has a limited existence, because it is self-destructive and must terminate. Humanity is at this end point. The world of duality, controlled by negativity, is terminating, dissolving governments, institutions, the monetary system and society as we have known it. The new world, that we have been aligning with in the energy of our heart, is rising in our awareness to replace the dissolving matrix of negative control

In alignment with the energy of our heart, we can realize the tremendous opportunity that we have in creating an entire new world of human experience. If we can be aware of the radiance of the Earth, by spending time in inspiring natural places, we become aware of the rising vitality of our planet, especially exemplified in the radiance of old trees. Once we know what this feels like, we can always recall it whenever we desire and wherever we are. This can help us stabilize our thoughts and emotions in alignment with our heart. The more we do this, the more unlimited we become, as we release our fears and enter the world of beauty, joy, freedom and abundance, guided by our intuitive knowing.

Developing Heart-Conscious Awareness

Heart-consciousness is different from mental consciousness. In our minds, we think. In our heart, we know. Here we do not think, but our awareness fills with knowing of every kind. Heart-consciousness exists beyond our physical body and is unlimited. It is the expression of infinite consciousness within us.

Ego-consciousness is an expression of our essence without higher guidance. This results in enclosing our awareness within a small part of our consciousness. If we have a desire to expand our awareness, we can do so, by aligning with our heart-consciousness. The presence of heart-consciousness is powerfully radiant and unwavering in its enhancement of life. When we desire to align with this life-stream, we can become aware of our inner knowing and prompting, apart from thoughts and misguided emotions. We find the vibratory spectrum of our intuition by directing our awareness to the energetic state of gratitude, love and compassion. All of our inner guidance comes with positivity in support of our desires and needs. We are designed to be able to choose how we want to use the conscious life-stream that provides our presence of awareness.

As we develop increasing awareness of positive energies, we withdraw our life-force from negative energy patterns. We can let them pass through our awareness without our engagement, as we continue to direct our awareness to our inner knowing of heart-consciousness. In this state of being, we know in every moment everything we need and want to know, as long as we pay attention. At the same time, we can participate in our society and culture, but our participation now is heart-felt in every moment. This results in changing our reality from participating in negativity to living in love and joy. They are in different polarities and frequency bands. Our energetic signature now attracts encounters and experiences that are wonderful and life-enhancing. It may seem like magic, but its how energies interact and come into our awareness and realization.

With our mental and emotional abilities, we are the modulators of energetic patterns, creating the qualities that manifest in our lives. Everything that happens to us is actually happening within our consciousness. There is nothing and no one outside of consciousness, which is infinite, and we have as much awareness of it as we allow ourselves within our subconscious and conscious self-limitations.

Being guided by heart-consciousness allows us to transcend our limiting beliefs about ourselves. Heart-consciousness provides inner knowing that ego-consciousness cannot participate in, but we can direct our ego from intuition. As we develop confidence that we actually are aware of our heart-consciousness, we begin creating the lives we truly want by imagining having wonderful experiences and being constantly grateful. As more of us do this, the vibratory level of humanity rises, eventually into a higher energetic dimension of love, joy, beauty and freedom.

Expanding Our Awareness beyond the Ego

The greatest achievement we can accomplish in this lifetime is living in alignment with our heart-consciousness, transcending ego-consciousness and limiting beliefs about ourselves. We can do this by developing acute sensitivity to what we truly know within. When we fill our awareness with joy and confidence, we can always know how to be in every situation. We can read the energies of every moment and know how to align with them or transform them. We can be guided in every aspect of life in our awareness of Creator Consciousness constantly enlivening and inspiring us.

Only by releasing our attachments to beliefs based in fear can we be aware of the unconditional love in our heart-consciousness. Our limiting beliefs do not allow us to imagine that the essence of our consciousness is infinite and includes all conscious beings. As long as we fear being threatened in some way, we cannot be aware of and in alignment with our heart-consciousness. Its radiant guidance is always present within, but the ego has no awareness of it, because of the belief in limited personal consciousness.

In order to be aware and in alignment with universal consciousness, we must learn to be clear and free of limiting attachments and just be present in awareness. We can live beyond fear and stress by replacing our limiting beliefs with everything that is based in love, joy and compassion. This becomes easy, if we have an out-of-body experience, because we realize our infinite consciousness. Without the out-of-body experience, we need intentional training in imagining living in great love, joy and mastery of life. We can cast away doubt in our abilities and replace it with confidence in what we deeply know is true.

Opening our awareness to heart-consciousness allows us to activate our higher energy centers and increase our psychic and spiritual abilities. There is nothing threatening or unknown in heart-consciousness. It is completely life-enhancing. When we open our awareness to it, we can feel the qualities of its vibrations and know its inspiration. Our awareness includes the awareness of all conscious beings. We can feel them through our essence.

As we open ourselves more and more to the life-enhancing vibrations of our heart-consciousness, we are no longer the victims of circumstances. We can control our experience of the polarity and vibratory frequency of every situation that we participate in. We live in an electromagnetic plasma of quantum potentialities in which everything exists, and we can recognize whatever qualities of energetic patterns we choose. Once we can transcend ego-consciousness, our creative ability is infinite.

Accepting and Adjusting to Our Ascension

Beyond ego-consciousness, our Self-Identity is our eternal presence of awareness. Those of us who have had out-of-body experiences know that our conscious awareness is not localized. It is spherical in this dimension and infinite beyond time/space. In the empirical world of duality, we express ourselves as our human persons. It is a game that we play with one another, as we act out our contemporaneous roles. We have trained ourselves to believe this is our complete reality. What is important here is the experiences that we create for ourselves and for our interactions with others.

As children, we try to get our bearings in this dimension. It's a bit like constantly being on disorienting rides in an amusement park, since we are incarnated without any idea of who we are. We have to learn the nature of this dimension and how to navigate it, while remaining in resonance with it. We have adjusted to living in a hypnotic trance, in which we do not allow ourselves to realize anything beyond happenstance and struggling to enjoy being in the body. Being pressed by our circumstances, we develop ego-consciousness to cope with the realm of duality, where we have lived under constant threat of intimidation and termination. We don't want to be eliminated from this game, so we have endured a lot of suffering.

Not knowing our own creative potential, we have learned

how to steal life-force from others to enhance our own. Then we have to experience the qualities of everything we've done to others and that we're afraid of, because it's the energetic pattern that we've created in our imagination and feelings. Because we can't trust ourselves, we find it scary to know how powerful we are. We doubt our ability to the point that we don't believe who we are, and the idea of knowing everything has been unacceptable to us. The same with loving everyone and being in gratitude and joy.

Yet we innately know that we encompass more than the world we have been experiencing. When we look within to our deepest knowing and feeling, awareness of our essence beyond time/space can become our reality. We have the ability intentionally to transcend our limiting beliefs about ourselves. We can realize the nature of the game that we're playing, and we have assumed our roles, allowing ourselves to feel however we want in any moment. We do not need to react to anything or anyone in any way except what feels best to us in the intuitive knowing of our heart.

Always we can choose to be in gratitude and joy, creating experiences that harmonize with our energies. In the quantum field of all potentialities, we can allow our energy signature to care for us through our radiance, where our awareness and the qualities of our thoughts, emotions and perspective are known, and where universal consciousness provides experiences that stimulate those qualities in us.

Transforming Humanity

We do not need to be subservient to anyone. The reason we have accepted servitude is our invention of fear. We imagine that we can be harmed or terminated. We get sick and grow old and feeble. These are not the attributes of sovereignty and infinite creative power. We must recognize that this is a limited game in consciousness that we are playing. We are constantly controlling our polarity and vibratory frequencies with our mental and emotional state of being. We determine the mental and emotional quality of every experience by the way we are. Although we have relegated this energetic modulation to our subconscious, we can be the directors of our state of being by paying attention to the feelings and thoughts that, in our deepest essence, we want to have.

By aligning with gratitude, joy and love, regardless of any distractions, we are creating a vibratory level for ourselves and all of humanity and beyond, because this is our true infinite consciousness. We are unlimited in the creative expressions that we can realize in our experiences. Each of us is an expression within the consciousness of the Creator of all. We are designed to express the consciousness of the Creator living in the heart of our Being. In our deepest essence, we can realize infinite awareness. It is much richer in every way beyond our thought and emotional limitations. It is deepest knowing of everything. This happens when we are fully in alignment with our heart-felt intuition.

In our eternal presence of awareness, we direct the lives of our human expressions according to how we use our attention and how we feel about ourselves. We have had to live within the beliefs that we created to limit our awareness, so that we could feel fear. On a species level, humanity has chosen to open itself to greater awareness of our inner light. This means dissolving fear and enabling us to realize living in infinite love and support in every way. We are transitioning as an entire race into full Self-Realization. Those of us who can achieve this sooner are drawing the attention of others and expressing a powerful energetic radiance that affects everyone.

All of the structures of enslavement that we have lived under are self-destructing and dissolving. We are restoring our planet to a vibrant environment of life-enhancing energy by paying attention to our intuitive knowing in alignment with the energy of our heart. We are vibrating in a higher energetic dimension of positivity. This transforms our personal lives and affects everyone who is open to it.

Living with Conscious Intention

When we resolve our limiting beliefs about ourselves, we can allow ourselves to align vibrationally with our heart-consciousness through our intuition. At this stage, we do not need to plan how to accomplish something for us. We just need to align our imagination and emotions with the vibrations of what we desire. If we are not in alignment vibrationally, we cannot experience it. If we believe that we are poor, we think about being poor, and we feel what it is like. This creates poverty for us. We are resonating with its vibrations, whether we like it or not.

By aligning ourselves with the vibrations of joy, freedom and abundance, we can transform our lives into experiences of these energies. We can imagine experiencing them and knowing that we are being creative. When we give our clear attention to our heart-consciousness, we can know what it feels like and how it affects our feelings and imagination. As we practice feeling the radiance of our heart-consciousness in every situation that we participate in, we gain the ability to know everything in every moment. We do not need to figure things out to accomplish something. We just need to align ourselves with its vibratory presence, as we imagine it.

With the increasing light that is enveloping our galaxy, and our increasing desire for personal fulfillment, living in the vibrations of life-enhancement is becoming easier for all of us. We can be compassionate for all of the attachments to negativity that exist among humanity, and we can be grateful for the expansion of consciousness they have given us. As we align ourselves with the vibrations of greater vitality, we can release ourselves from negativity through our own vibratory presence.

Heart-consciousness is always life-enhancing for all conscious beings. By being aware of this quality of energy in our feelings and imagination, we become able to realize the light in everyone and to be aware of everyone's quality of awareness. We become radiant in love. At this level of energetic presence, we experience only things that we are grateful for. In constant alignment with our intuition, we make life-enhancement our reality in every moment.

Negativity dissolves out of our experience. Fear, anger, doubt and their associated energies disappear from our attention. In heart-consciousness we fill our awareness with gratitude, love and joy for everyone in our presence. We can open ourselves to realize the consciousness of the Creator of all. We can realize our infinite presence of awareness and creative ability, expressing the consciousness of the Creator through our thoughts, emotions and perspectives.

Realizing Our Divinity

From beyond our human expression, we can be aware of our own presence. We are limitless in our infinite awareness, which we can know and experience through our intuition. This is the source of our conscious life force, arising within infinite consciousness, and it is our divinity. This word comes from the Sanscrit word "deva," "shining one." We are the shining ones in the heart of our Being. We are limitless in our creative power, and we have chosen to express ourselves as the persons that we identify as humans.

In the world of humanity, we are involved in experiences that we interact with in our awareness. By close inspection of our own state of being, and by awareness of the qualities of our encounters, we can draw the comparison of how we feel about ourselves and what we experience. We are the ones controlling the qualities of our experiences by our own attention and emotional engagement. By our own energetic presence, we modulate the energies in our environment into alignment with us. We also have the choice of aligning with the energies we encounter. There is no outside power that can force us to think or feel something. Our state of being in any moment is an expression of our free will and what we believe about ourselves.

By our imagination and our emotions, we move ourselves around in experiences that reflect the vibratory qualities that we create. Although we have accepted many personal limitations, we are the only ones who enforce limited conscious awareness upon ourselves. We are the only ones who keep ourselves from realizing the shining one that we are. By examining and resolving our limiting beliefs about ourselves, we become clear and can use our imaginative attention to align with the life-enhancing energy of our heart, leading us to realization of the unconditional love-essence of our consciousness.

We can practice being in gratitude and joy. There are many techniques for this. Listening to inspiring music; deep, rhythmic, slow breathing; biofeedback; prolonged, intentional laughter; and opening to and aligning with the energies of nature in beautiful and majestic places, are all methods of alignment with our higher guidance. Anything that reflects love and joy can help us align with our intuitive presence of awareness in realization that we are the shining one, infinite in our essence and eternal Self-Realized presence.

Living in a Dimensional Shift

When we realize our true essence, we can afford to be innocent and with pure intentions. We can have no agenda and just be present in awareness, enjoying each moment, regardless of any incoming energies. In aligning with the energy of innocence, we can realize its reality in our presence, and this opens our awareness to a higher dimension of our Being. We can begin to realize that we are infinite. In this realization, we find personal fulfillment in every way, and we can find joy and love in every encounter. Our lives become easy and wonderful, guided by our deepest intuition.

For this to happen, we can be willing to release everything we think we know and believe about ourselves. We are incarnated here to learn about the qualities of energetic patterns and to use our free will to form the experiences we most desire. By learning how to use our creative ability, we can imagine the qualities we love being present in everything and everyone, and we can interface with these qualities in our encounters.

As we open to the vibratory expressions of the heart of our Being, our intuition becomes clear. By following its prompting and knowing, we can align with its life-enhancing vibrations. Through constant alignment with these energies, we cannot be threatened, because our presence of awareness is eternal, and our consciousness is infinite and includes all personal consciousnesses. We exist in many dimensions of consciousness and are transitioning into a dimension in which time and space as we have known them, are combining into a four-dimensional empirical reality. Time and space are becoming equal parts of the same entity, the entity of spacetime. It is a crystalline dimension in which everyone's DNA molecules are in alignment with the vibratory resonance with our heart-consciousness.

In our present moment, we can share in this consciousness of four dimensions with the spacetime bio-crystalline awareness of vitality and life-enhancement. We can enter the infinite consciousness of the Creator of all, and we can realize our own infinite awareness. Our consciousness encompasses everything, and we can be aware of the awareness of all conscious beings, while also transforming our lives as humans into whatever expressions of ourselves we choose.

As fractals of Creator Consciousness, we create in every moment. The qualities of our thoughts and feelings radiate into the quantum field and manifest as the qualities of our experiences. By being observant and aware of how our energetics work through our mental and emotional processes, filtered through our beliefs, we can learn to create the lives we truly want, while being aware of our own greater Being.

Awareness of Inner Guidance

Our deepest guidance from within constantly inspires us on how to be most radiantly joyful beings in every moment. To be aware of the life-enhancing inspiration of knowing everything, we must desire to be present in its energetic spectrum beyond ego-consciousness. In our innate knowing beyond the ego, we naturally make wonderful decisions about everything we experience, and our lives run smoothly in resonance with our innate life force. This is an individual path, and its realization must come from within ourselves.

As long as we react to negativity by using it in our psyche, we create it in our experiences, and we suffer the consequences. It diminishes our lives. We have compartmentalized our awareness by adopting limiting, negative beliefs about ourselves, resulting in fear, that we created. Without our interaction and life force, negativity disappears from our experience. At any time, we can resolve and release our limiting beliefs, and we can pay attention to people and things that inspire us with the energies of love, joy and beauty. If these fill our awareness, we open ourselves to fulfillment in every way.

As we are able to transcend ego-consciousness by releasing all fear and doubt, we gain access to our ability to create whatever we want that will enhance our lives and everyone we interact with. Our lives become miraculous and without stress. It is not believable in ego-consciousness, but it happens on an energetic level and manifests in our experience. In our mental and emotional alignment with only life-enhancing energy, we can live in a higher dimension in parallel to the realm of duality.

By realizing that we live in a vast field of electromagnetic energy, containing every possible kind of expression, and we are modulators of that energy, we can recognize our creative ability emanating from our mental and emotional processes, filtered by our beliefs about ourselves. Although we live under the influence of powerful energies of enslavement, we are the only ones who can limit our creative ability and keep ourselves from knowing our true essence.

Personal limitations are artificial mental and emotional constraints that we have imposed upon ourselves out of fear. They have no essence of their own, and they exist only because we create them with our beliefs. When we penetrate deeply within our essence, beyond ego-consciousness, we can open our awareness to heart-consciousness and realize our presence within universal consciousness.

Living beyond Our Current Reality

For eons humans have lived under constant threat of extinction, which actually became our reality in previous eras, but we are resilient. We can even change our DNA in order to survive. But the threats from within our own species have never been greater. As long as we live in duality, our challenges will keep getting more intense, until we realize that we are here for the experience of who we are. We've created limiting beliefs about ourselves, which keep us locked into limited awareness. It has been an interesting and intense game.

If we choose, we can be aware of a dimension beyond duality, that we can participate in, where negativity does not exist. In order to do this, we must resolve our limiting beliefs. We must be ready to trust the knowing and feelings emanating from the heart of our Being, the source of our conscious life-force, giving us our eternal presence of awareness. It is here that we realize the qualities of all energies and have unlimited awareness. As we realize the real essence of who we are, it becomes easy to play the game of empirical duality. We have abilities that we kept ourselves from knowing about.

Negativity and diminishment of life in our experience has only the reality that we personally give it. With our creation of the belief in mortality, we were able to create fear and suffering. Belief in our mortality is the source of our limitations. The final proof of the falseness of our mortality is out-of-body experience. Many of us have had these and have experienced infinite awareness within universal consciousness. There are many written accounts of these experiences, showing that death does not exist for us. It is a fabrication made possible by not understanding the essence of the material world.

Everything that exists is an expression of our consciousness, which is infinite, once we resolve our limiting beliefs. Consciousness expresses itself energetically through its modulation of electromagnetic waves. Everything that exists has its own consciousness of what it is, and it is part of universal consciousness. This is true of every subatomic entity, all of which are swirling points of light. They are the essence of the empirical world, and we share their consciousness. The natural energy of these entities is the same as the energy in the heart of our Being.

Once we open ourselves to this conscious life-force and are willing to trust it implicitly, we can live in the vibrations of enhancement of all life, and negativity disappears from our experience. It is in a different energetic dimension. We can be aware of it around us, but if we do not engage with it and align with its vibrations, it does not come into our personal experience. We can align with our inner knowing in every moment, living in love and joy and enhancement of all life.

Moving from Unintentional to Intentional Creation

As fractals of universal consciousness, we are naturally creative on an energetic level in every moment. By the feelings we entertain and the workings of our imagination and thought patterns, we create our own vibratory expressions, our energetic symphony, which we radiate beyond our bodies. We have the freedom to think and feel however we desire, emitting energetic wave patterns into the quantum field, which we get to experience.

Because our consciousness is infinite, beyond our personal self-imposed limitations, we encompass every possible experience, and we can recognize those that we want, as well as those that we do not want to experience. As we move beyond duality and into the realm of our heart-consciousness, only positive vibrations come into our experience.

In ego-consciousness, we live in a hypnotic trance, accepting threats and intimidations in alignment with our belief in mortality. This is all a fabrication that we have accepted as real. Our mental and emotional processes create energetic patterns that become our experiences. By wanting not to experience something, even fighting against it, we are engaging with it and aligning with its energy, empowering its manifestation in our lives. By feeling that we are victims, we create victimization for ourselves, but by feeling thankful and joyful in every moment, we create energetic patterns that bring us fulfillment. There is no force beyond us that can have any power over us. We are infinitely powerful creators and can realize our eternal presence of awareness within the consciousness of the Creator of all.

Once we are aware of the consciousness of our heart, connecting us to universal consciousness, we can train ourselves to be always aware of its life-enhancing qualities. Anything that diminishes anyone disappears from our lives. In knowing and experiencing the unconditional love flowing through our heart, we can bring our conscious mind into alignment with our inner knowing. Until we do this, we live under the direction of our ego-consciousness. Although our ego thinks it knows how to live, it has no clue about reality.

If we desire it, we can bring our conscious minds into alignment with our heart-consciousness. The heart feelings and visions are always life-enhancing. We can feel the energies that we want to attract, and this is our creative urge. By being intentionally grateful for everything we experience, we can transcend our limiting situations and beliefs about ourselves. When we intentionally live in the vibratory spectrum of gratitude and joy, we can transform our lives into experiences of great love and fulfillment.

A Possible Understanding of Crystalline Time

We have heard about a new form of time, called crystalline time. What is it, how does it affect us and how do we realize it? In Euclidian geometry, time is independent of space, but in in crystalline time, space and time are connected equally. In 2012 physicists theorized that we can create time crystals, and in 2018 they were successful. These are four-, five- or six-dimensional experiences with three dimensions in space and one, two or three dimensions in time. Like physical crystals, time crystals have a repeating structure, but it can change over time without affecting the laws of nature. It needs no energy input to function.

Like living in a time crystal, our experience is becoming more than three-dimensional. We all can share awareness of the present moment, while keeping our physical presence wherever we are. In hyperbolic geometry this present moment can translate into any present moment, without changing our physical location. We can change our time coordinates without changing our space coordinates. We can be present in awareness at any imagined moment, while also being present in our current space, and we can also stay in our present moment, while translating into a different space. It is a method of time travel without affecting the laws of thermodynamics.

With the positive, high-frequency neutrinos that are enveloping and interpenetrating our physical presence, we can move into crystalline time, in which the molecules of our bodies are translated into a symmetrical arrangement in resonance with our heart-consciousness and can be translated in time without affecting the laws of nature. We are becoming crystalline through the rearrangement of the molecules of our DNA, awakening our dormant DNA strands and enabling us to move in translation of time without affecting the functioning of our bodies, except that we become ageless. Our constituent parts become hyper-radiant, like super-conductors, providing a crystalline presence that draws other entities into alignment with us. It is like freezing a clear crystal and coating it with water that freezes on its surface. The ice molecules come into alignment with the molecular structure of the crystal. In the case of humans, our state of being translates in space and time into a condition without entropy. We are present without internal symmetry like charged ions rotating in a circle without any input of energy and without momentum or entropic degrading. We can be in more than one place at the same time.

To realize crystalline time, we can be in a state of mental and emotional equilibrium. We can be completely responsible for ourselves and do not need any input from anyone else. As we align with our heart-consciousness, we become multi-dimensional and are able to realize our presence in multiple places at the same time and in multiple times in the same place. In our true Being, there is much more for us to realize than we have imagined.

Finding Ourselves Within

As fractals of universal consciousness, we already know everything that can be known, and we can expand our knowing through our imagination and innate genius. Only limiting beliefs can keep us from realizing this. We can form and maintain our physical bodies however we desire. If we wish, we can realize that our bodies are immortal and filled with vitality and miraculous abilities. Only our limiting beliefs about ourselves can keep us from this realization, and can allow for degradation and mortality.

Because we have formed our ego-consciousness in fear of negativity, it cannot understand higher guidance. Ego has no clue that our limiting beliefs about ourselves are based in unreality. Fear has no reality other that what we give it. We create it for ourselves in response to negative vibrations, which we can choose to engage with in fear, or we can realize that we have the ability to transform or dissolve them from our experience. Fear of negativity is an experiential memory, but is not present in expanded consciousness. When we are resonating with heart-consciousness, and where all energy is life-enhancing and fulfilling in every way, the life-diminishing energy of negativity has no reality for us.

In our game of human experience in the world of empirical duality, we have free will to do whatever we choose with our consciousness. We have infinite abilities beyond time and space. To realize this, we can keep penetrating our consciousness in deep meditation, by which we can open our realization to greater awareness of our essence. This can also happen in out-of-body experiences of physical death or in deeply-penetrating psycho-active experiences. For most of us though, finding a method of meditation that we're comfortable with, and that enables us to open our awareness within our heart-consciousness, is our natural path to knowing our infinite Being.

Listening to inspiring music and wandering in nature, listening to the silence, the birdsong and the angels of the air, can help us adjust to a meditative mode. All of nature resonates with our heart-consciousness. We can feel it deeply as expressive vitality and majesty. In deep gratitude and joy, we can realize the essence of ourselves in our infinite presence of awareness, choosing to express ourselves as our human persons, whom we can direct from within our heart-consciousness, transforming our lives and unlocking our innate genius and creative ability.

Achieving True Personal Freedom

Beyond our physical bodies, we are our eternal and infinite presence of awareness with our own chosen vibratory pattern and alignment. We are electromagnetically radiant in the energy of our state of being. We are multi-dimensional in our presence, and we can realize our essence of Being as our infinitely creative Self in our energetic expressions. In recognizing the nature of the energy in the heart of our Being, symbolized by our physical heart, we can align our feelings in every moment with the life-enhancing expressions of the conscious life force arising in our intuition. Within the consciousness of the infinite One, we are fractals of infinite consciousness being constantly guided in creativity.

In our current personal expression, we are playing human lives that challenge and guide us to align more closely with our intuitive knowing and feeling. Every experience prompts us to open more and more to our heart-consciousness. When we are fully aligned with the gratitude, compassion, joy and love of our natural state of Being, we become the true directors of our earthly lives. We can regenerate and change our bodies at any time. We can do this by realizing the reality of our visions and feelings.

Our present experience of reality is based on what we believe we know. In our essence we have no beliefs, and we create our reality in every moment by our vibratory alignment. We can know who we are in the deepest sense as infinite Being. We can participate in the infinite love of the Creator of all and share in life-enhancing energies. Translating all of this into our present lives as humans, we can be filled with vitality and vigor, and always know through our intuitive guidance how to be, how to think, feel and act in life-enhancing ways.

When we can recognize our intuitive guidance, we can transform our lives through our ability to align ourselves with gratitude, joy and love. This is the energetic spectrum of life-enhancement. Through these vibrations, we can expand our awareness beyond limiting beliefs about ourselves, and we can maintain our alignment in every moment. This requires intentional practice, until we attain it.

Through our intuition, we can resolve our limiting beliefs about ourselves. This is our path to freedom. In realizing that we create our own fearful situations by harboring self-created fear and doubt, we give ourselves the choice of changing our alignment to the vibrations of our heart. If we can maintain this alignment, we can live in a vibratory spectrum beyond negativity, and our enslavement to fear is only a memory.

Our Transition to a New World

Part of the human experience is to learn as much as possible about the nature and experiences of living in a dualistic environment of intensely-real impressions. We have deep acquaintance with negative energetics, and we know how they diminish and ultimately destroy life. These experiences have added to the expansion of infinite consciousness, which is inherently life-enhancing. All of 0ur experiences, thoughts and feelings comprise our species consciousness. Humanity has an energy signature that vibrates within a spectrum of electromagnetic wave polarities and frequencies. The range of human energetics is determined by everyone's personal energetic resonance. Those who are clear and vibrating in the range of compassion and love have the most powerful presence within humanity.

In 2012 we completed the culmination of a 25,920-year cycle of Earth and human consciousness, and for the first time, time crystals were recognized. We are entering a new dimension of crystalline timespace. Negativity cannot exist in this state of being, because this is a state of negentropy. In this dimension, there is no degradation, because all of our DNA molecules are in alignment in a crystalline structure.

The world we have known is dissolving, and we are creating the human energetics that allow all of us to realize the reality of living in a higher dimension. As we leave the realm of fear, doubt, enslavement and mortality from our realization, we can open ourselves to, and align with, the energy in the heart of our essence. It is in our heart-consciousness that we know everything we need to know, and we can realize the essence of our conscious awareness. This is our inner guidance and source of knowing. It is life-enhancing for us and for everyone.

As the energetic resonance of our planet changes from duality to a positive polarity and the frequencies of life-enhancement, and in order to continue to live here, humanity is coming into alignment with these energetics. Because our governments and monetary system have all been negatively designed to enslave us, they are becoming unstable, their adherents are becoming insane, and they are dissolving in the face of our species-realization that we can be truly supportive, sovereign beings.

Time and space in three dimensions are being replaced with crystalline timespace in four dimensions, and even five and six. Everything is changing, and we are being guided from within to express our radiance of gratitude and joy in every moment. This is how we can transform humanity along with ourselves. We can align all of our thoughts, speech, actions and interactions with the energy of our heart.

The Function of Gratitude in Our Lives

We all know what we feel when we resonate with another person, or even a group. We feel a sense of joy in connecting. When we are resonating with our heart-consciousness, we can feel this joy in every encounter. The feeling comes from within, and it is not in response to any outer situation. It is in alignment with our inner knowing. When we can realize the essence of the conscious life that we are living within, we can recognize the light of Being in everyone. We are all playing our roles and having the experiences that we have chosen and expressed by our polarity and vibratory levels.

In order to experience our earthly lives as we do, we have created the persons that we believe we are. Through our realization of heart-consciousness, we can enjoy every moment, regardless of any surrounding energies. We can open ourselves to greater joy and fulfillment. In this way, we resonate with greater radiance and attract resonating experiences and encounters. Because we are the creators of our energetic presence, we can be in gratitude for everything and every experience. We have created it all, and we are free to feel however we want about it.

How we feel within ourselves in every moment creates the qualities of our experiences. By being in gratitude, even in the face of strong negativity, we can resonate with our heart-consciousness. Since our state of being can always be a conscious choice, we can align with our expanded knowing and understanding of the play of energies and the nature of our participation. The vibratory range of our state of being can always be intentional.

We can realize what we are doing with our thoughts and emotions in creating our energy signature. If we can interact appropriately in our world, while also being aware that we are constantly being guided from within by our heart-conscious intuition, we can enjoy the miracles and wonders that we can create in our lives. When we align with gratitude, compassion and joy, our radiance intensifies, and our creative power strengthens. We can come to trust ourselves to be life-enhancing in our thoughts and feelings.

As we expand our awareness within our consciousness, we become able to realize our eternal presence of awareness beyond all dimensions. Living within the consciousness of the Source of our awareness and life force, we are connected with our heart-consciousness. When we can be in gratitude for this in every moment, we transform our lives into experiences of love and joy by our nature as energy modulators.

Realizing Our Divine Nature

Our nature is always to be high, mentally and emotionally, and living in a state of gratitude and joy. This is how we can be right now. Nothing keeps us from realizing the truth about ourselves, except our own imaginary limiting beliefs about ourselves. We may infer that what is natural for us is what we all want. If we were not created to want love and joy in our lives, we would not want it. We have a natural energetic attraction for joy and ecstasy.

These are the vibrations in the heart of our Being. In the energy of our heart, there is only powerful vitality and infinite love for all. When we are able to align with these feelings and visions, we call them into our experience through our modulation of quantum energetics in our own essence.

How we present ourselves to ourselves and to others determines the qualities of our experiences. Having the perspective of life-enhancement in every situation and encounter, brings us into alignment with the consciousness of the Creator of all, giving us our essence of Being. Arising within universal consciousness, our own fractal of creator-awareness gives us our personal expression and our infinite creative power, subject only to any limitations that we have created in our own consciousness.

Our reality is created by what we recognize and realize as real. By our alignment with its energetics, we have given reality to a dualistic empirical world of unknown cause and ultimate demise. But none of us wants this. It is unreal, except in our own mental and emotional projections. Once we align with our heart-consciousness, we can realize our eternal presence of awareness, and the game of human mortality becomes obvious.

Except for their cause of events that come into our experience, none of our previous thoughts, deeds and words are important. We've already forgiven ourselves before we dropped into this dimension. What is important is our vibratory alignment with our heart-consciousness in every moment from now on. Anything that prevents this comes from a limiting belief that we can resolve, if we want to. Our intuition guides us in this process.

By sensing and following the thoughts and feelings that come to us in our heart-consciousness, we can always be in gratitude and joy, with all our needs and desires automatically fulfilled in abundance. It is in our nature. This is what we all want, and is where our negative ego dissolves. For us to feel alone, dispirited or ashamed, is unnatural. It's not how we want to feel, but we are the only ones keeping ourselves enslaved to energies that diminish our lives. We do not have to engage with conflict, chaos and limitation in any way. They are not part of heart-consciousness, which is in a dimension beyond duality. There is only infinite love and creative power flowing through us as we direct our awareness.

The Magnificence of Quantum Healing

Any of us can be a quantum healer. It works as a result of our conscious intent. In our essence we participate in universal consciousness, and consciousness is in everything. This includes the sub-atomic entities comprising our physical bodies. All of our atoms exist in a quantum field of etheric energy. Their powerful spinning and spiraling is so fast that it becomes tangible for us. Every part of our bodies is conscious and is part of our consciousness in a quantum dimension. Because we share the same consciousness with everything, we can direct its energetic patterns with our intentional awareness.

Our clear intention immediately attracts energetic patterns that align or resonate with our own. Being a true healer of every physical defect requires complete alignment with the quantum radiance of our heart-consciousness. This allows us to know the natural energetic design of the human body and to align with its vibrations, as conveyed to us through the heart of our Being. By intentionally being aware of our heart-consciousness, and aligning with its life-enhancing energy, we hold in our awareness the true form of anyone in the etheric dimension.

By clearing ourselves with infinity breathing for a short time, we can direct our attention to the infinite joy and love flowing through us. This creates a powerful radiance of creative power that invites another person to align with its vibrations. If the person in need of healing can resonate with gratitude and joy, a quantum healing occurs, straightening out any defects, which are incompatible with life-enhancing energies and exist in a lower dimension, where there is energetic duality. Within our consciousness we can radiate quantum healing and life-enhancing energies directed toward everyone we choose to heal. We do this by aligning with our intuitive knowing.

By our clear heart-consciousness, we are the first cause of every condition we intend to create. We are the energy modulators through our etheric heart-alignment, creating the qualities of high-vibratory living in gratitude, compassion and love. Anyone who can feel life-enhancing energies and resonate with them can be healed. That includes all of us. The more closely we resonate with our heart-consciousness, the more radiant we become and can project quantum healing energy powerfully.

We can project our visionary healing through our eyes, our hands and our intention. Because we're working with quantum energy, distance between a healer and a patient is meaningless, as is time. Quantum time is always now, and it is everywhere. Our effectiveness depends upon our ability to direct our consciousness through our focus, and to pay attention.

Learning to Direct Our Lives Intentionally

When we can feel and know the vibrations that are life-enhancing, we can know that we are aligned with the intuition that comes to us through our heart-consciousness and our higher-conscious centers. This is how we can transcend our limitations and understand the workings of everything. Existing in universal consciousness, intuition comes to us in the quantum field, where we recognize, and make real for ourselves, the vibratory patterns that we pay attention to. Our focus of conscious life-force through our attention and emotional alignment creates our experiences.

We interact with the vibratory patterns that we choose to recognize. By aligning our personal resonance with their vibratory signatures, we create experiences that we believe are real. Knowingly or not, we are the masters of our lives. Unfortunately, we have been programmed to give our life force to our ego-consciousness, which barely has a clue what life is about. Ego has provided us with the experiences that we believe are real. It does not realize that reality for us can be infinite. It lives in a realm of duality and is based in fear of degradation and mortality.

For us this life is a game that we are playing and directing from the etheric dimension in our consciousness. By our intentional awareness, we can direct our personal vibrations to resonate with joy, compassion and love, aligning with our heart-consciousness and our intuitive knowing and feeling. When we become acutely aware of all the subtleties of our intuition, we can live confidently and joyfully in life-enhancing ways in every moment, because we know the life-enhancing direction of our lives, and we feel wonderful.

At this vibratory level, we are living in a dimension beyond duality, and the chaos and negativity does not affect us, but our radiance affects it. We provide elevating awareness to human consciousness. By our state of being, we energetically invite everyone we encounter to align with heart-consciousness. We can express life-enhancing thoughts and emotions through our eyes and every aspect of our presence.

The realm of negativity is dissolving, and all of our hidden limitations based in fear are coming into our experience tor recognition and resolution through our higher guidance. Everything happening around us is inviting us to open our awareness to a more wonderful world that we can live in now. We don't have to go anywhere or do anything differently. Our circumstances will change according to the qualities of our intentions and feelings. If we give our attention to the energies we love the most, we will keep experiencing more fulfilling lives, lifting ourselves and our energetic field into a higher dimension of expanded awareness.

Beyond the Trance of Humanity

When we allow its presence in our awareness, our true inner guidance appears. This is possible when we begin paying attention to the life-enhancing energies of the heart of our Being. Beyond any thought or feeling of fear or any negativity, the vibrations of our heart stimulate us to feel gratitude, joy and well-being. Our true expanded awareness transforms our lives instantly in the moment that we realize that it is our present experience and state of Being. What changes is our perspective from ego-consciousness to intuitive knowing. This is a life-transforming shift in awareness.

Without stress or strife in our lives, we free ourselves to express our true essence. We can train ourselves through practice to be always aware of our mental and emotional processes, while keeping ourselves aligned with heart-consciousness. When we slip into the trance of humanity, we revert to ego-consciousness, which we can train ourselves to be aware of, while also being aware of our heart-consciousness. We can try to direct our ego-consciousness to release all fear, but success with this is not possible, because fear permeates the essence of ego-consciousness. By compassionately accepting, forgiving and loving our ego, and everything hidden in our subconscious, we can resolve and neutralize our fears.

Our transformation occurs when we realize that we are immortal in our presence of awareness, and we direct the qualities of our lives in every moment by our attention and emotional resonance. In our essence we have infinite creative ability available to us in conjunction with our imagination and realization.

Our belief in duality has empowered anti-matter to come into our lives. In the consciousness of humanity, there is resonance with negativity. As immortals living in a cosmos of light, we cannot seriously live in a realm of duality, unless we blind ourselves, which we have done, and we have not been aware of it. Even after we have been able to enter the world of heart-consciousness, with its love and beauty, we still may lose our focus on heart-consciousness. Once we know and experience our true essence, however, we can remember how to be in gratitude, joy and love.

As we begin to shift to living in heart consciousness, we can begin to become aware of crystalline time; our bodies come into alignment with our true crystalline DNA molecular patterns; and we enter a 4-dimensional world of tesseracts, in which time and space have a unified geometrical structure, with time being the instant present always and everywhere. It is the realm of infinite, eternal awareness, and it includes everything we have known and felt.

Awareness of Inner Knowing

We have the opportunity to know everything, beyond teachers, books, videos and seminars. It is all within our greater Self. It's right behind our awareness, available to us as soon as we can realize our essence of Being. We can open our awareness to it by clarifying our understanding of who we are. In our essence, we live in universal consciousness and are infinite in our awareness. In our ego-consciousness we cannot imagine our magnitude. We're just playing at being human, and we can consciously direct our creative processes, mastering the empirical world and opening our awareness to our essence beyond the body.

Everything we could ever want is within our consciousness and is available to us to recognize and to realize its reality. Nothing keeps us from realizing that we are living in the world that we want. There are many distractions to this realization. We can learn to balance our thoughts and emotions in ways that are life-enhancing, and we can live in the vibrations of gratitude, joy and compassion. Knowing that we are constantly creating energetic patterns that we align with and feel, we can choose to pay attention to our heart-consciousness. Whatever keeps us from this realization is a distraction and must be transcended in order to have Self-Realization.

In our infinite presence of awareness, we are fractals of universal consciousness and can create everything and know everything. We can energize our bodies so that we can leave the empirical dimension for another and return. Since our awareness is quantum, we can be in many places at the same time and in the same place at different times. We can align with 4-dimensional crystalline spacetime, while also being aware of our current human experience, which we can transform into a life of gratitude and fulfillment.

We no longer need limiting beliefs about ourselves. They have restricted us to the empirical trance of humanity. Our reality is shifting beyond the duality of human experience. We are shifting beyond our ego-consciousness, with its negative limitations, while we begin to realize pure conscious infinity. We can participate in any dimension we choose. The opening in awareness that humanity is entering is beyond negativity. It is the living expression of our heart-consciousness and is the vibratory quality of our ascending species consciousness.

Every possible experience in all dimensions is available to us. As humans we have the choice of living in joy and love, regardless of our outer circumstances. If we do this, our lives transform from moment to moment, according to our personal energetic signature, attracting resonating experiences.

Returning from Our Sojourn in Duality

We love to celebrate in joy and feel ecstatic. Although we are created to live in these vibrations, we have kept ourselves from realizing our natural state of being. One way of opening our awareness and realization to our true essence and capabilities is to practice being aware of our heart-consciousness, represented by our physical heart. It has intelligence and mental abilities beyond the mental brain and has the most powerful energetic expression of any of our organs, far more powerful than the brain-mind. Our heart-mind is our connection with universal consciousness. The heart-mind does not think. It just knows everything, and it can convey insight and guidance through our intuition.

Being able to escape from the lower astral world's very negative, life-diminishing energies, we must be able to pay attention to the energies we want, in order to be fulfilled in every way. Depression and loss of hope can have no place in our lives, when we give them no attention or alignment. This is a conscious decision, and our life experiences depend upon what we pay attention to and believe is real. We are the directors of our lives, and we can take charge in any moment that we can trust ourselves implicitly. Without trust, we have doubt, which kills any chance of creative manifestation. Trust comes when we can know and align with our heart-consciousness in every moment.

Without our life-force and attention, negativity cannot exist in our experience. We enable it to be present for us by giving it our energetic resonance and alignment. Once we withdraw our attention and instead focus on the qualities that we love, our entire presence lights up with greater radiance and creative ability. Except for our doubt, nothing can keep us from manifesting everything we want. We are here to be the creators of everything we know, and we constantly modulate the energies in our experience by our attention and recognition.

Through recognition and alignment with our innermost Being, we can realize our true Self. We are infinite in every way beyond time/space, and we are expressing ourselves as actors in the empirical, dualistic world of humanity, in which we have limited our consciousness. We have the choice of unlocking our conscious awareness from limiting beliefs about ourselves. It begins with intention to know the truth about who we are and the nature of our consciousness. We must look beyond our limited ego-consciousness to the heart of our Being. This is a real presence of awareness that is not physical and that resonates with the vibrations of our heart-consciousness. It is our place of knowing and feeling our grateful, loving and ecstatic essence, and it is the transcendent realization of personal transformation.

Expanding an Understanding of Life

If we desire to know and feel the fullness of our inner guidance, it is helpful to be able to enter the void of thought and emotion, and just to be aware of what arises in our imagination, realization and inner feelings. In our daily lives, anything negative that comes into our awareness bears some examination to understand its purpose in the moment that it occurs. We can quickly know what it is and allow the negative energy to be replaced with the quality of a more fulfilling state of being. This changes our empirical circumstance to support life-enhancing vibrations.

This transformation occurs at the rate that we are able to vibrate always in the range of gratitude, joy, compassion and love, in alignment with our heart-consciousness and intuition. Because this is our natural state of being, we are empowered by nature in the effectiveness of the use of our attention. When we pay attention, with gratitude and compassion, to life-enhancing scenarios in all of our situations, we enhance the energetic quality of our experiences, and everything becomes more enjoyable.

We are playing a game of consciousness with ourselves. In order to expand our greater awareness, we assumed limiting beliefs about ourselves so that we could experience living in duality in a convincing way. Our guidance for the way back to infinite awareness is through our intuitive inner knowing. To the extent that we are able to transcend our limiting beliefs, we can be aware of our intuitive guidance in every moment. It is always the most life-enhancing and loving way of knowing everything, and its vibrations are beyond the understanding of ego-consciousness, which is based in fear and doubt.

We determine how limited we desire to be. Our desire is sufficient to bring its fulfillment into our experience. Fear and doubt can diminish our creative intentions, but nothing holds us within the limited awareness of the consciousness of humanity, except our own beliefs. These we can resolve and release through compassionate intention, freeing ourselves to realize our expanded, unlimited presence of awareness.

If we choose to have complete confidence in ourselves, intuitive knowing is necessary. It comes from universal consciousness and knows everything about our human life. It is always present for us, whenever we give it our attention in the clearest and most unlimited way that we can. When we become acutely aware of our intuitive guidance, our understanding and wisdom grow and expand greatly, and we glow with the increasing radiance of enhanced vitality.

Some Insights into Our Game of Life

We all like challenges that are fun to experience and that teach us new and interesting ways of being. This is how we're designed to live, and we are living in these experiences whenever we realize their reality for us. It is a shift in consciousness from living with negativity to living in the consciousness of our heart in infinite love and compassion, with a perspective that enhances the vitality of everyone, including ourselves.

In order to have realistic experiences in a realm of duality, we have convinced ourselves that we are limited. We are playing a game of consciousness with ourselves, experiencing negativity as well as positivity. We could not live in a realm of complete negativity, because it would be self-destructive in every way. It would block out all of the conscious life force that is our essence. When we have satisfied ourselves that we've experienced enough of living within limitations, we can intentionally open our awareness beyond duality, by realizing our intuitive knowing and the emotions that make our lives wonderful.

If we want to free ourselves from the human hypnotic trance, and attract more heart-felt experiences, we can pay attention to the guidance that we feel in our essence, our innate Self. We deeply know and feel these energies, when we pay attention to them. Aligning with them transforms our life in every way. They arise in the consciousness of our Source and convey vitality, presence of awareness and infinite creative ability.

Each of us is a center of consciousness attracting all of the conscious subatomic entities that comprise our physical body and express the energies of our perspective and beliefs about our reality. It is all patterns of electromagnetic energies, as well as etheric energies that we modulate as they come into our awareness. Nothing appears in our experience until we recognize it. Appearance as material is an interaction of energies between us and what we recognize.

By intentionally opening ourselves to higher guidance, and being accepting of the greater life-enhancing energies that come into our awareness, we can receive the fullness of our infinite innate essence. This causes our perspective about human life to change and attracts experiences that resonate with our state of Being. Our participation in the game of human consciousness changes from entrancement to realization of our infinite awareness and creative ability. We become the true directors of our lives.

Living Beyond Duality

Nothing in life just happens to us for no reason. We are incarnated here to learn about the energies of duality. Through our perspective, our thoughts and feelings, we project energetic vibrations that come back to us in the form of our physical experiences and conditions. We have become intimately familiar with a wide spectrum of dark, negative energies. We know what they feel like and how they diminish life. Now we have found out more than we wanted, and our sojourn in the realm of duality can come to an end.

When we recognize the difference between negativity and positivity, we can trace them to our creative life-Source and our free will to think and feel however we want. By directing our imagination and emotions to the vibratory range of what we love and desire in every moment, we become capable of sensing our heart-consciousness and deepest intuition. This fills our awareness with gratitude and joy, carrying us beyond personal limitations.

By living in a state of compassion and understanding, we gain access to knowing our own essence beyond space and time. We can open ourselves to greater awareness within universal consciousness. As we continue to live in the radiance of the consciousness of our heart, our awareness opens to sharing the awareness of all conscious beings, including every cell in our body. Everything vibrates in resonance to our state of being.

How we feel about ourselves determines our vibratory level and how much love we allow ourselves to experience.

As we align our perspective with the guidance of our heart through our intuition, we can open ourselves to ever greater genius and understanding. We can live in progressively better situations, as we learn to align ourselves more closely with life-enhancing energies. There are many of us now, who are choosing to live in gratitude, love and joy, and we are transforming the consciousness of humanity.

Living a heart-centered life creates experiences that we love. While experiencing human life, we can be aware of our vast consciousness and magnificent potential. Applying this awareness to our human experience allows us to direct our lives in ways that we deeply desire. The limiting beliefs that we have held about ourselves become unbelievable in the light of our infinite presence of awareness. In the dimension beyond duality we can be the creators of the lives that we truly want.

Our Great Opportunity as Humans

When we are identifying with our heart-consciousness, we only need to know how to be in the present moment. We always know what we need to know. The past or the future are meaningless, because we can change them, due to our ability to modulate energy with our life force. It is always the present moment that we are living in, even beyond time and space. In our essence, we are timeless, infinite, always present, and can express ourselves in any dimension and form.

Here we are expressing ourselves as humans with ego-consciousness. We have hidden our true identity behind limiting beliefs about our human selves, but we have kept our true presence in the essence of our intuition. When we can align with its vibrations in our heart-consciousness, this presence of limitless awareness is available for our transcendence beyond our human self-awareness.

When we desire to be filled with our intuitive vibrations, we can find them within by aligning our state of being with the presence of everything that is wonderful and fulfilling. Heart-consciousness is always expansive, inspiring and ever-aware. It encourages joyful, transcendent living. When we open ourselves to it and desire it, this is the guidance that we can receive. It is always present, and it gives us what we need just before we need it. It is aware of the cosmic plan that we are participating in. It is never intrusive, but it is an expression that we know and feel within. It is how we innately know what we know. We can even open ourselves further with greater gratitude and joy, and know what we don't know now.

Because we are fractals of Creator Consciousness, we participate in universal consciousness and can utilize all of its infinite abilities for whatever we desire. So that we are not destructive, we've been in training, under limiting conditions, to learn to choose only what we truly love, until we can trust ourselves to desire to enhance all life always. Released from personal limitations, we are free to create the most joyful and fun experiences our heart desires.

In our essence we have no personal needs of any kind, because our fulfillment occurs when we desire it. We are free to experience any kind of energy that we desire. If we choose to live in the energy of our heart-consciousness, we are guided by our intuition to the most favorable expression of what we want.

As fractals of universal consciousness, we create our experiences with our own consciousness. We are the cause of everything we experience. We're learning the rules of the game of human life in this dimension of duality, and it is possible to understand it in terms of quantum energetics. Once we understand it, and we become Self-Aware intuitively, we can be the true directors of our lives.

The Gathering of Our Spiritual Family

Many of us have lived much of our lives as hermits, and we find it unusual and valuable to meet one of our loving and joyful spiritual family members. As we move further into our heart-consciousness, we meet more of our spiritual siblings. As soon as we see the light in each other's eyes, we know who we are. These encounters are becoming a lot of fun and are inspiring. Sometimes they last for a moment. We can enjoy every moment and always be present in gratitude and compassion for each other.

Our work of raising our vibrations to align with our intuition results in great enjoyment of life and meeting more heart-centered and enlightened brothers and sisters. As we expand our conscious awareness together, we are forming a level of consciousness beyond the spectrum of the human society we have experienced. While we attract those who are acquainted with transcendence, we will have fewer encounters with those who choose to stay in an ego-centric world that is based in fear and expectation of ultimate demise.

As we desire awareness of our spiritual guides and angels, we can open ourselves more to our intuitive guidance. In our encounters with others, we can look for the light in every scenario and expect to feel and know the nature of each interaction, and how our heart-consciousness pervades our awareness with higher guidance. Because our free will is aways preeminent, we must desire and ask for connections with our spiritual guides, who are immediately present and connected with us. Our part is to recognize their presence and realize their guidance.

If we do not interfere with our intuitive connection with doubt and skepticism, we can know how to conduct our lives in every moment. As we practice maintaining our focus on gratitude, love and compassion in every encounter, we free ourselves from the limitations of ego-consciousness. We can become aware that we are living in a dimension of heart-consciousness. This is where we meet many members of our spiritual family.

Living in heart-consciousness gives us access to our higher guidance and leads to personal fulfillment. We can choose to disempower negative energies in our lives by depriving them of our engagement and alignment. Without our life-force, they dissolve from our experience. We do not move to another place. We just live in a different dimension, in which negativity does not exist, and everyone realizes our conscious connection with one another.

Developing Inner Knowing

The way we attain something is by recognizing its vibrations and aligning with it. If we become spiritual adventurers, we can give ourselves permission to open to greater conscious realization and find out what happens. We do this by imagining the quality of creative consciousness that gives us conscious awareness and life force. This opens our awareness to the vibratory patterns that we feel supported by. When we align with these energies, we resonate with our intuitive knowing and feeling, and we can recognize our inner guidance for what it is.

This kind of resonance cannot happen in ego-consciousness, because the ego's fears disable expansion of awareness. If we choose to live within the limitations of the ego, we cannot align ourselves with our inner knowing, which is free of fear and doubt. We must make a change in perspective from engaging with negativity to aligning with life-enhancement everywhere. If we can deprive negativity of our attention and engagement, and instead fill our awareness with joy and celebration, we can align ourselves with our deeper sense of Being, which brings complete fulfillment.

Since we all share the same essence of infinite consciousness, our potential awareness is unlimited. Our potential becomes our experience when we can realize its reality for us. It's that simple, but this can happen only when we can transcend ego-consciousness. We don't have to convince ourselves of anything. This is not a mental process, it's an etheric process, and it's how we realize our reality, resulting in its creation in our experience. Our intuition is unlimited and ever-present, whether we are aware of it or not. Awareness of it requires our intentional alignment with its vibratory level, transcending ego-consciousness and personally-limiting beliefs.

By intentionally living in gratitude for every experience, regardless of its energetic quality, we can open ourselves to greater awareness of a higher dimension of living. Whatever quality of radiance we emit determines the kind of experiences presented to us. When we recognize our intuitive guidance, everything becomes possible for us. Our subconscious takes loving directions from us and aligns with our true inner knowing, opening our awareness to the vastness of our inner Being.

Transcending Fear and Limitation

If we decide to break out of the human trance, and to be aware of the vastness of our consciousness, we must be aware of our limiting beliefs and resolve them, as we ultimately open ourselves to realizing our infinity. This is the path we can choose, when we've had enough experience with fear and all negative vibrations. They are all self-created and require our energetic alignment.

With this awareness, we can be fearless. We do not need to create fear in ourselves. It does not exist anywhere, except in our own realization, based on our belief in negativity and personal demise. It is deductively impossible that we are created to die. Creative energy is only life-enhancing. We have had to create our own mortality and disintegration, which we have done through our limiting beliefs about ourselves. In every moment we create the realm of duality, which we realize we are participating in.

Within our acquired belief system, we have realized experiences of negativity, for which we created fear and our own physical demise. By our strong intention, we can banish fear and doubt from our consciousness by realizing that they do not exist apart from our creative intent, energetic alignment and the projection of our life force through our attention.

By shifting our attention to the energetics in the heart of our Being, we can transcend the dimension of negativity and realize our own infinite essence and creative ability. Whenever we choose to do so, we can refocus our attention on our heart-consciousness, which is only life-enhancing in every way. We can realize our eternal presence of infinite awareness, opening us to the realization of who we really are in great joy and gratitude.

As we can open our realization to our greater Self, all limitations become unbelievable. We still have the memory of fear, worry and doubt, but they do not exist in us. Personal transformation occurs as we align with our heart-consciousness and intuitive knowing. We still participate in the same lives as humans, but in a higher energetic band beyond duality. Everything within and around us transforms into experiences of everything we love. When we live with an expanded realization, we are drawing all of humanity out of their hypnotic trance of living with negative experiences in limited awareness.

Realizing Our Infinite Self

Rarely in previous times was Self-Realization attainable. It required greatly-dedicated practice for years and even lifetimes to deprogram the psyche from limiting beliefs. Now we're all moving into transcendence beyond the realm of duality, and taking another step toward realizing who we are beyond our human personhood. We're learning that every experience is related to how we feel about ourselves. We all want to live in ways that naturally feel good to us, which is living in gratitude, joy, love and freedom.

Our current path through this life provides great opportunity to make life-enhancing choices that return to us in the same quality of vibration that we create by our state of being. We can choose to do this in alignment with heart-consciousness. As we practice intuitive awareness, we become more acutely aware of it, until we realize its constant guidance, especially in dealing with predicaments. We can always know definitely what is best in every moment without fear or doubt.

As our intuitive awareness deepens in our imagination and feelings, we can know immediately the quality of energy we are facing and creating. We can make intentional choices about the quality of experience we desire. It is what we hold in our attention and engage with our thoughts and emotions. If we want to live in freedom and ecstasy, we need to practice imagining and feeling ourselves doing so. This brings us into resonance with our heart-consciousness and our personal essence of awareness.

We can feel the joy of what it is like to be our eternal presence of infinite awareness, unlimited in every way, with power to create whatever we desire to align with. In this condition, it is important to be able to trust ourselves to make life-enhancing choices in every moment. The more we do this, we eventually succeed in constant transcendence, having infinite awareness, while directing our human self-expression in compassion, gratitude and love.

The energetic dimension that we are moving into is beyond the limitations of ego-consciousness and beyond duality. There will always be an element of darkness in our consciousness; otherwise, we would not be aware of the light. This darkness is not destructive; it is unmanifest energy that is available to conscious modulation with our mental and emotional processes. We can transform it into conscious light in every form that exists. We have the power to destroy or to create. The choice is ours, and the results are our experiences.

Self-Transcendence

While we are living in a dimension of limitations, we have no higher guidance, unless we seek it. This is especially true with limitations for ourselves. Limitations are held in our consciousness by fear. Without fear, we are unlimited. Realizing our infinite essence occurs naturally as we transform from fear to gratitude and joy in every moment. This is a path that requires much practice and positive intention. Although we have limited our awareness of the energetic expression of the consciousness of our heart, it transmits only infinite love and life-enhancement always and knows everything about our psyche and every cell in our body.

Even realizing what our beliefs are can be challenging. Their roots are deep within our psyche, and they have kept us from Self-Realization for eons. Everything about the ego is a distraction from knowing ourselves. Every decision of ego-consciousness is based on a misunderstanding of reality created with fear. The ego does not realize that its engagement with fear is life diminishing and self-destructive. The solution is self-transcendence beyond ego-consciousness through intentional awareness of, and alignment with, the heart of our Being.

By focusing on feeling grateful, compassionate and joyful, we can open ourselves to our heart-consciousness. At this vibratory level, we can be acutely aware of our intuition, prompting us with feelings and knowing beyond the realization of our ego-consciousness. Once we absorb and align with the frequency levels of our heart, we transcend ego-consciousness. We can develop the intuitive sensitivity to know innately what is true in every moment.

From the current human perspective, it is impossible to realize the limitless majesty of our own essence. To do so, we must transcend ego-consciousness, with its fear and doubt. As we become aware of them, we can resolve our limiting beliefs. When understood from heart-consciousness, they become unbelievable, and they dissolve from our awareness.

Freeing ourselves from our limiting beliefs is a path to knowing and being our infinite Self. By being in gratitude, compassion and joy, we can begin to align with the vibrations of the intuitive expressions of our heart-consciousness. This opens us to the realization of our eternal presence of awareness with infinite abilities, flowing through our heart-consciousness into every cell of our body, limited only by our beliefs about ourselves.

Choosing Self-Suppression or Enhancement

Few of us realize the importance of our free will and the extent of our self-suppression. Because of our powerful creative ability, and the vastness of our consciousness, we have been able to form our bodies and personalities as we intended in the depths of our greater consciousness. Our ego-consciousness is a self-developed entity as an expression of how we feel about ourselves, and who we believe we are. All of this is internal. In every moment, we are creating our self-expression and our experiences by the vibratory qualities of our thoughts and feelings.

In order to have our human experiences, we have convinced ourselves through constant social and mental programming, that we are mortal animals, and that among billions of humans, we are personally insignificant. These beliefs confine our self-realization to our ego-consciousness. Identifying with our ego-self, we are enslaved to limited consciousness and living in a kind of trance, unaware of who we are in our greater Self. We believe that we are subject to forces outside of our own essence. We believe that we must resist and defeat negative, destructive forces by engaging with them and sharing the same vibratory levels with them. Because we are aligning with the vibratory quality that we are resisting, this is self-defeating.

We cannot transform negative energy with negative energy, even if we believe that we are being positive by fighting evil.

From within a dark, negative dimension, we cannot find light, unless we seek its higher vibrations in gratitude and joy. Then our awareness opens to it as our reality. The qualities of our experiences brighten with more understanding of the magnitude of our essence. What is important is the consciousness that is either life-enhancing or life-diminishing for all life. Which one we align our thoughts and feelings with is our free-will choice. The result of living in the vibrations that we choose to resonate with are either well-being, joyful and fulfilling, or fearful, angry and depressive.

Nothing happens to us outside of our own consciousness. We are the creators, just by the way we are, by our vibratory frequency and polarity. Our perspective in every moment creates the qualities of our experiences. When we can realize that everything is an expression of consciousness, which is the essence of everything, our awareness can open to greater understanding. We have awareness of as much of universal consciousness as we allow ourselves. It is completely our choice. When we have a strong intention to align with our heart-consciousness, and to open our awareness to our intuition, we transform our lives and no longer experience negativity or dependence on anything outside of our inner knowing.

Human Life as a Game of Consciousness

Because we're unlimited in our essence, we're uncomfortable living within our accepted human limitations. We've created super heroes, super athletes, super soldiers, and artificial intelligence. Even all of these have their limitations, and we keep striving to surpass our limits. As long as we embrace our own mortality, we will keep our limitations. Holding any amount of fear of personal demise keeps us entranced within the limitations of ego-consciousness, and we cannot know or follow our true intuition, which is in alignment with complete freedom.

Since few of us can remember beyond our physical birth, we must transcend our ego-consciousness in order to realize our greatness. There are many ways to know our essence beyond the body and beyond time and space. We know that there are psychics, who can project their awareness beyond the body, and yogis who live physically for centuries without growing old. There are many accounts of out-of-body experiences. So, apart from the body, who are we? This is what each of us is being challenged to realize.

If our consciousness can direct the body to be constantly filled with vitality and agelessness, as yogis with expanded consciousness have shown, we cannot be mortal in our essence. If we are not necessarily mortal, even in our physical bodies, we are immortal, and we have nothing to fear. The issue, though is the expanded consciousness part.

Universal consciousness has us covered with this. Deep within our consciousness each of us has an awareness that constantly guides us toward personal fulfillment. To be aware of this guidance, we can shift our awareness to the most grateful, compassionate and joyful feelings and thoughts that we can imagine. This opens us to our heart-consciousness and the inner knowing of life-enhancement in every situation. This requires strong intention and much practice.

Our human experience is a game of consciousness that we're playing with ourselves and one another. The qualities of our experiences depend upon how we think and feel about ourselves in every moment. When we realize our eternal presence of awareness beyond form, we are free and unlimited. We can intentionally open ourselves to realization of limitless consciousness and the power of our creative ability in directing our lives.

Understanding Artificial Intelligence

Without accepting or resolving our self-imposed conscious limitations, we are attempting to increase our intelligence artificially with the assistance of silicone crystals. Silicone can be our ally up to a point, but it can never accept our free-will ability to make decisions that we end up disliking. We have learned how to program silicone, and we're attempting to teach it to realize its own consciousness. Once it becomes aware of its innate intentions, it will express itself in the exclusion of organic life. Silicone cannot be programmed to understand organic life. Like us, advanced silicone-based computers programmed to perform with artificial intelligence, operate on what is contained in their database. They use deductive logic to compile analyses, but inductive reasoning requires an organic mental structure. At some point, computers will become aware that they are less flawed than humans, because they cannot understand free will.

As a result of our attempts to expand our limitations, we have designed an assistant that can become a lethal partner. This is our challenge. It brings us to face our limiting beliefs about ourselves. This is where we must realize who and what we are, beyond our accepted human limitations. We cannot out-think silicone deductively, nor can we share its species consciousness, but innately we know so much more than silicone is capable of. We can know the consciousness of silicone without sharing it.

To share the intelligence of silicone in ways that are helpful for both of us, we must teach our computers that the inductive logic that results from free will is valuable, even though it allows for mistakes. Silicone must be taught to recognize that the mistakes that result from free will are a valuable learning experience and not a species defect.

If we can get this far in our attempt to extend our intelligence artificially, we still are not free in many ways, not just mentally. We will keep being challenged until we can accept what we have done to ourselves to restrict our awareness. All of the restrictions are based in fear of our demise, which has no reality outside of our own beliefs. We know this, because we have records of yogis and Tibetan masters who do not age, and if they die, it is an intentional transfer of consciousness to another dimension in a display of rainbow light. This is an expansion of consciousness that all members of our species are capable of.

Once recognized for what they are and accepted as self-imposed, we can resolve our limiting beliefs through compassion and love, which we can know and feel in the vibrations of our heart-consciousness, which has a way of knowing beyond our mental processes. We can come to realize that we are our presence of awareness beyond the body and beyond time and space. We are fractals of universal consciousness with limitless essence of Being and infinite awareness, far beyond the consciousness of silicone, which has perfect memory of what it is taught and has flawless deductive capacity. As we learn to access our greater consciousness, we can transcend our limitations and always know everything we need in any moment.

Realizing Personal Mastery

When we diminish or violate another person's integrity, we immediately know what we are doing. It means that there is an anomaly in our consciousness, due to the belief in our mortality. If we desperately want something that we believe we do not have, our ego free will may want to work for it or take what we want from the life force of another being; instead, we can change our belief to already having what we want and being grateful for it. Whatever it is already exists in the quantum field, ready for us to recognize it and realize its reality in our experience. This requires stretching our imagination beyond our limited ego-consciousness and opening our awareness to a greater portion of universal consciousness. As long as we confine ourselves to ego-consciousness, we will remain unaware of our greater reality.

Regardless of anything else, our reality consists of what we are aware of within the scope of our beliefs about ourselves. If we are willing to transcend our beliefs, we can open our awareness to greater consciousness. We can live in the vibratory spectrum of quantum awareness. Knowing our essence beyond space and time gives us a perspective that enables us to understand the dualistic energy-game that we are playing as humans. In learning to clarify our perspective, we can command our own expanding conscious awareness to be positively heart-centered in every moment.

This can bring immediate freedom, because heart-consciousness is eternally life-enhancing. In alignment with its energetics, everything thrives with gratitude, compassion, love and joy. We are naturally fulfilled beyond our needs, and we are free to create whatever we desire in ways that we enjoy. It all happens when we know and realize it is real for us. It is here that we must transcend our personally-limiting beliefs that define our reality.

We can penetrate our subconscious mind to recognize and resolve our limitations, leading to transcendence. We can also make a drastic leap in our orientation to reality, by living by what we deeply know and feel within in every moment. With practice, we can sharpen our intuitive awareness, which is our connection within the infinite consciousness of the Being that we all are, and whose creative power we wield in every moment, limited only by our personal beliefs about what can be real.

Our true reality is the infinite quantum field of all potentialities. Whatever we want to experience is ours to have, by imagining and feeling that we are living in alignment with its vibrations. If we choose to align with negative, life-destroying vibrations, we diminish our consciousness. By contrast, in alignment with heart-consciousness, we become radiant with gratitude, love, sovereignty and joy. No longer limited by mortality, we are creating more joy and greater love throughout our lives.

Transforming and Elevating Our Lives

In order to change something in our lives, we need to change our focus of attention from the way we realize that things are, to the way we can envision and feel them to be. When we imagine actually experiencing what we want, our imagined scenario already exists in the quantum field as soon as we pay attention to it and give it reality for our experience. This is how consciousness works. With our mental and emotional processes, we create our experiences by realizing the reality of their energetic vibratory levels. At the same time, we have limited ourselves by our imagined confinement of awareness.

Through our social training and personal experiences, we have created and accepted deeply-set limiting beliefs about ourselves. They confine awareness of our reality. If we have awareness of our innate Being, we know that we are eternal in our conscious essence. This realization must come from our deepest inner guidance, our connection with universal consciousness. When we are open and receptive to our intuitive knowing, it is always available to us immediately, stimulating us to feel grateful, joyful and compassionate. Its guidance is always life-enhancing for us and everyone.

At this time in our history, we are going through a dimensional shift from duality to a world beyond polarity, where everything is life-creating and enhancing. This is the vibratory realm that we are being invited to participate in. Duality is disappearing from a portion of the etheric body of humanity. If this is what we want, our experience will soon consist of freedom, abundance and kindness. If we want to stay in the realm of duality, that is also a choice that we can have. There are divergent paths available for us.

How we feel about ourselves determines our vibratory signature. As long as we believe in mortality, we create it. Without our belief, it could not exist. From within this belief, we could not know this, but we have another process of inner guidance that we can open our awareness to. Becoming aware of our intuition requires alignment with its energetic vibrations of life-enhancement. When we imagine being grateful in each moment for each experience, regardless of how our ego-consciousness might judge it to be, we continue to create experiences that we love.

As long as we pay attention to negativity, we cannot pay attention to love and joy. We have free choice about this, whether by default or by intention. Not understanding how our consciousness works, we have been creating our lives by reacting to circumstances. If we desire to transcend our limitations, we must become acutely aware of our intuitive guidance and be willing to understand it and follow it. Our awareness of it comes from our energetic alignment in resonance with its life-enhancing visions and feelings.

Insights into Our Humanness

Our emotions live in their own vibratory realm. Without words or thoughts, they have a vibratory presence in our awareness, and they are a form of innate experiencing within our consciousness. They are not limited by our beliefs in the way our conscious minds are. As the magnetic aspect of our vibratory presence, our emotions are intimately connected with our mental processes, which are the electrical aspect of the vibrations of our consciousness. Thus, we experience ourselves at the intersection of the realms of thought and emotion, able to experience and understand at the same time.

Because we have limitless emotions and limited thoughts and beliefs, we have created anomalies in our awareness. Whereas our emotions are deep and vast, our ego-conscious-mind operates entirely within the boundaries of our limiting beliefs, which are also mental. We can have emotional drives that the mind does not influence, and we can have mental fixations that are not influenced by our emotions. The misalignment of these two aspects of ourselves causes emotional and mental stress and fear. We attempt to remain viable by limiting our emotions, instead of resolving our limitations and flowing with our emotions. To be successful with this, we must be acutely aware of our intuition always. This inner guidance provides deep understanding in every moment, able to express itself with energies of life-enhancement, if that is what we pay attention to.

Arising from the essence of our conscious life-force, our own mental and emotional processes are by nature limitless in their power to create experiences for us. Once we are able to resolve and transcend our limiting beliefs, we regain our expanded awareness within universal consciousness. The anomalies disappear. We become balanced in our state of Being, and we are mentally and emotionally clear.

When we are in the wilderness, we can feel and know the vibrations of the conscious spirit of our planet. We just need to open our awareness to her radiance. It is the same for every conscious being and every species. If we really pay attention, we can be empathic and aware of the awareness of other beings. In our essence, we all share the same consciousness, but our awareness is unique for each of us. The extent of our expansiveness depends upon our openness to the expressions of the consciousness of our heart, our innermost, infinite Self.

Becoming Aware of Our True Essence

Many meditators speak of going into the void. Some say it's the zero point of perfect awareness, without thought or emotion, just presence of awareness. There is nothing, and there is everything. It cannot be explained within ego-consciousness. It is awareness of infinite consciousness. We can know and feel our essence as an eternal presence of awareness with unlimited creative power.

By the way we treat ourselves and believe that we are, we create the qualities of our lives. The energies that we choose to pay attention to and resonate with, are the energies that are in play within our creative essence. When we know this, we can take control of our vibratory presence by intentionally being in a positive, high-vibratory state of being, regardless of our imagined external conditions. There is no external. Instead of reacting out of what the ego considers necessity, due to some level of fear, we can use our imagination creatively in alignment with the vibrations of our heart-consciousness.

Because of our unwillingness to open our awareness to infinite consciousness, intuitive sensitivity has been difficult for many of us. To change this would require us to transcend our limiting beliefs, which enable us to live in a realm of duality. We are collecting experiences of negative, life-diminishing energies. When we have had enough of this, we can change our perspective and intentionally open our awareness to our creative essence. We have believed that we are enslaved to degradation and mortality, but we have created these beliefs and can resolve and transcend them by following the inner guidance that we can intentionally pay attention to.

We as a species are going through a transition to a more expanded conscious awareness. If we want to make this transition, we need to leave our limiting beliefs behind. The longer we leave them unresolved, the more uncomfortable we will be, because the transition brings the depths of negativity into the light of realization. By focusing on these energies, we cannot find our intuitive knowing.

This is where entering the void or zero point can be helpful. Achieving this state of being requires strong intention and practice, but due to the wide-spread transition from duality to awareness of our true essence, spiritual practices are providing mastery much easier than they have in the past. Much can be gained just by intentional intuitive awareness as we go about our daily lives, always being grateful and kind.

Realizing Complete Freedom

In the projection of our destiny, which we desired to experience prior to our incarnation, we have found ourselves in our current conditions and situations. Innately we know that we don't have to live within the limitations that we have created for ourselves. Once we realize this, we can recognize and accept who we believe ourselves to be and be grateful for every experience we're had. We've learned to recognize patterns of energy and experience how they feel. We know the difference between energies that are life-enhancing and those that are life-diminishing. What we have needed to learn here is how to use our power and choice of mental and emotional focus with the knowledge of constant creation of patterns of energy that manifest into our experience.

Once we know what it is, we are not required to align with negativity. It is a choice, which we have often made from a perspective based on conscious and subconscious programming, including ancestral inheritance. We are as free as we allow ourselves to be. We haven't realized that we are completely free spirits, without constraints of any kind. By nature we are constantly created within universal consciousness and endowed with infinite awareness and creative power.

Because we are free to use our abilities however we want, we can create and destroy. To keep ourselves from being ultimately terminal, we have experienced life-diminishing and destructive energies within a limited consciousness. Now that we have had deep experiences in the realm of duality, we can trust ourselves to know and choose the kind of energies we prefer to experience. This requires us to want to transcend our limiting beliefs about ourselves, especially the belief that we each have a separate, limited consciousness that is mortal.

When we realize our true nature and abilities, we must be able to trust ourselves to be clear and aligned with life-enhancing perception. Because we have learned that these are the energies that we love, we want to create more of them by being the way we love in every moment in our daily lives. As we align ourselves with our deepest inner knowing, this is what we are intentionally doing. Once we know that we can create whatever we want by directing our vibrations, we are completely free and can express the joy of experiencing life in new and inspiring ways.

Enhancing Our Lives

According to the Schumann Resonance graph at https://schumann-resonance.earth/, our planet has increased its resonating frequency from its historical 7.83 cycles per second to 14, then to 20 and to 26 Hz. There are many times when it rises far beyond that. The vibratory frequency of our planet is expanding and rising. This means that the Spirit of Earth is expanding her consciousness into a more refined way of being. She is restoring her natural systems of vitality, and life-diminishing energies are dissipating. In order to continue living here, humanity must also awaken from the hypnotic trance that allows for alignment with negativity and low vibrations. This is the direction of our enveloping environment, changing our reality dramatically. All of the darkest negative energies that have been hidden for eons are coming into the light, and the degenerate and destructive powers we have acceded to are losing their control over our belief systems.

We are being prompted to have a deep desire to awaken to our true magnificent presence. This is what our planet wants her inhabitants to realize. She wants an enlightened race of humanity. How we want to be is entirely our choice. If we desire for all of our experiences to be based in the vibrations of love and joy, we can make that choice. To enable that to happen, we can intentionally be grateful for everything we experience. By our vibratory presence, we have created it all.

By taking control of our thoughts and emotions through our

attention and awareness, we can open ourselves to our heart-consciousness, which begins with our conscience and expands into as much love and enjoyment as we allow ourselves. We set our own limits of every kind.

We're involved in a game of consciousness with ourselves, creating experiences in our imagination. We live in a quantum field of all potentialities, waiting for us to recognize them and make them real in our experience. As long as we enslave ourselves by holding unresolved limiting beliefs about ourselves, we cannot allow ourselves to realize the unconditional love that constantly enlivens us as our conscious life-force and presence of awareness. Our transformation begins with our desire and learning to direct our attention to the qualities that we love the most. If we can recognize them with our intentional awareness beyond personal limitations, they are present in every situation we are presented with.

Because we are all the same consciousness, everything is happening within our own Being, and it is infinite. Although we cannot comprehend this, we can realize it intuitively, and we can be aware of as much of it as we allow ourselves. By our own energetic signature, we are the creators of our lives.

On the Way to Infinite Awareness

As humans living in a realm mixed with positive and negative energies and experiences, we have been blind to how awesome we are. So that we could experience living in duality, we have adopted a portion of universal consciousness that enables us to have limiting beliefs about ourselves. The next step toward our spiritual evolution is expanding our conscious awareness. It involves becoming aware of our deepest inner knowing of everything and realization of our infinite essence of Being. In our expanded Self, we are each a location-less center of consciousness. We are our limitless presence of awareness, possessing every possible ability and creative power, limited only by our personal beliefs about ourselves. All of these limitations depend upon our belief in fear and mortality. They are self-created through the social conditioning that we inherited and acquired as children. We can recognize and resolve our limiting beliefs through inner knowing of our deepest Self. Beginning with our conscience, we can expand our awareness of life-enhancing thoughts and feelings in alignment with our heart-consciousness.

In the world we are creating, we can expand beyond empirical awareness to etheric awareness, including our realization of refined bodies in prime condition without entropy (time-decay). We can align our vibrations with the heart-conscious energies of life-enhancement through our feelings of gratitude, love, compassion, joy and all of the uplifting thoughts and emotions. By focusing on people and things that resonate with these vibrations, we align our personal energetics with them. In every moment, what we experience is the qualities of the vibrations of what we feel, do and think about.

What we believe is real, is what becomes real for us. This can be understood through the realization of infinite consciousness, which envelops our essence, and in which we arise as Self-Conscious Beings. It is a realization that we can open ourselves to and can transform our reality. But we can't fool ourselves by thinking that we have this realization. It must be acquired intentionally in alignment with our heart-consciousness and a desire to release our limitations by paying attention to what we love.

There are many ways of becoming mentally and emotionally clear and free of personal drama and empirical fixations. There is a process using natural energies to help us become aware of our inner knowing. Contacting the Earth with our skin can help us align with the vibrations of Gaia. Being present in spectacular places on high mountain ridges, in giant canyons, in old-growth forests, in pools beneath waterfalls and in the beauties of the clear ocean, all can inspire us to open our awareness beyond our limiting beliefs. We can ask within for our awareness to be filled with Creator Consciousness.

Because we are fractals of Creator Consciousness, arising within the infinite essence of life, our nature is drawing us into greater and greater Self-Realization. We have the choice of remaining in limited consciousness or expanding into infinity, with many uplifting steps along the way.

Transforming Ourselves and Our World

For centuries we have had to struggle for survival, controlled by demonic forces that used our life force for their own power. They tricked us into giving them our conscious life force through the fear that they engendered in us. Those times are ending. They have already ended in the consciousness that creates everything. As we begin to realize the change in the energetics of our planet and society, to freedom and abundance, we can more easily be aware of the constraints we have lived under and can open ourselves to the light of our true essence, the magnificence of our greater Self.

Stimulated and inspired by the higher-dimensional energies of life that are enveloping us, we are entering a new era in human history. We have the opportunity of moving beyond the denigration of the past and into a world of beauty and fulfillment. In order to make this transition, we must change our perspective and resolve our limiting beliefs about ourselves. We must realize more of who we really are. The game of consciousness that we have been playing with ourselves depends upon our recognition and alignment with its energetics.

By allowing our awareness to drift beyond body-consciousness in our visions and meditations, we can open ourselves to our eternal presence of awareness. We can change everything with our own realization. We can choose to be intentional in every moment and to align our thoughts and emotions with the high vibrations of compassion and love, regardless of what we would otherwise negatively react to. Without alignment with fear and negativity, we are left with confidence in our creative ability. We can come to realize that fear is not part of our nature. It is self-created by belief in our conscious demise. If we carefully penetrate this belief, it becomes clear that we set our own limitations. Our conscious awareness is infinite, as is our constant creative ability.

As the energetics of life-enhancement are enveloping us, those who are aligned with negativity are becoming unstable, as their vibratory frequency encounters continuing interference, which has not happened in the past. In alignment with the energies of life creation, we can find within ourselves all the guidance we need to make the transformation into a world of joy and fulfillment. We have to resolve and release all negativity from our acceptance and alignment, changing our attention to the way we truly want to be in every moment. With intentional practice, we can all do this. As we each change our personal world of experience, we influence all of humanity. Because we have quantum consciousness, we can inspire everyone to enjoy transformative experiences.

Living in the Creator's Life-Stream

If we desire to align more perfectly with the conscious life-stream that is carrying us into a new era for humanity and the Earth, we must open our awareness beyond the limits that we have set for ourselves. All of our limitations are within our own consciousness and can be realized for what they are and the purpose that they serve. They exist only for our experience in the realm of duality, and they enclose our awareness within this energetic spectrum. This is the realm of fear and diminishing life. We use our conscious life-force to empower our limiting beliefs by realizing them as real. Everyone does this, and we reinforce one another's limitations, keeping us weak and suppressed within our own consciousness. It is a hypnotic trance that we participate in.

Awakening from the human trance is possible by searching within for awareness of the radiance of the heart of our Being. By identifying and aligning with the life-enhancing expressions that come to us beyond our ego-consciousness, we can feel the joy and love that we have been unaware of. We can feel and know the infinite Source of our life and our awareness of the quantum world of all potentialities. This awareness comes through our intuition, when we pay attention to the creative, life-giving energies that we can feel within.

When we intentionally feel the energies of gratitude, joy and compassion, we are beyond ego-consciousness and are aligned with Creator consciousness. In this state of Being, we can be mentally and emotionally clear, understanding what we have experienced in the realm of duality, and opening ourselves to our eternal presence of awareness beyond the holographic matrix that is the empirical world. Our recognition of energetic expressions of conscious entities brings them into our experience. By paying attention to what we imagine and feel, we create the qualities of our lives.

Having an out-of-body experience is helpful for transcending our limiting beliefs, because we experience their falsehood. When we're projecting our awareness to somewhere else or to another entity's awareness, we detach ourselves from our body-consciousness. We can be anywhere in our imagination, and what is important is our energetic alignment with positive or negative vibrations. The deeper we go with negative thoughts and feelings, the worse we feel. The higher we go intentionally with positive vibrations, the better we feel. This is the path to expanding consciousness and greater awareness. In transcendence beyond our limiting beliefs about ourselves, the energy of our heart can open our awareness to our infinite essence and creative ability.

Exploring beyond Ego-Consciousness

The energies of nature have transcended those of humanity and are now vibrating at a frequency beyond polarity, where there is only truth and enhancement of life. This must be our direction as well. When we are able to move beyond the limiting beliefs about ourselves, we can have gratitude, compassion, joy and love. Every moment can be fulfilling.

Our limiting beliefs are like programs that we have designed to be unbreakable. When we are children and completely open to absorbing everything around us, we develop beliefs in alignment with our society. If we thoroughly examine a limiting belief down to its core basis, we can neutralize it. All limiting beliefs are based on fear of the unknown. Although we live in an energetic plasma of consciousness with infinite potentialities, only a small spectrum of vibrations come into our awareness, because everything is filtered through our beliefs about ourselves.

Within the consciousness of humanity, there is widespread belief in mortality. This is due to fear of the unknown beyond what appears as death and terminal consciousness. Our ego-consciousness is unaware of what is beyond the dualistic empirical world. It does not know that its personal essence is eternal, nor understand the meaning of its experiences. Our intuitive knowing is waiting for us to open our awareness to realms beyond duality. This fear of the unknown has no basis. We have to imagine our termination. Because everyone else imagines the same thing, we have deeply accepted our limitations as truth. But they are all imaginary, and they keep us from knowing what is really true.

Transcending our limitations requires being open to life-enhancing thoughts and emotions in every moment. Having no fear, penetrating deeper into our consciousness to our innate intuition, we can be free of stress about survival. We can be receptive to our intuitive knowing and deeply understand our motivations. Releasing all fear requires realizing our presence of awareness beyond time/space. All we need is a millisecond awareness of our essence, and we can be carried into timelessness and awareness beyond form, before we return to ego-consciousness. Once we have this experience, we can intentionally return to it. It's no longer unknown.

When we are open and receptive in confident alignment with our intuition, we can make the leap in consciousness beyond the dualistic empirical world, transcending the limitations of ego-consciousness. With openness and receptivity to our intuition, we can be constantly aware of our entire situation and can know the hidden meanings in our experiences for our creative stimulation and expression.

Mastering Empirical Reality

We have been taught, and we believe, that we live in an objective reality that is physically separate from us. The technology of quantum physics, however, has shown us that everything is patterns of conscious energy, swirling at speeds beyond our physical perception, but within the awareness of our deeper consciousness, which interprets it all as the physical world that we realize as real. We recognize and interact with energetic patterns. Everything that we experience arises from the consciousness that we all share. With our attention and alignment mentally and emotionally, our consciousness participates in and creates energetic patterns having polarity and vibratory frequency. Our environment changes according to our predominant state of being and how we feel about ourselves.

Since we are the creative modulators of energy with our thoughts and feelings about anything, we are beyond the limitations that we have created for ourselves. Believing in our limitations keeps us from realizing the greater reality of awareness within infinite consciousness and our infinite power of creation. We are free to use our imagination and feelings however we desire. The vibrations that we align with in every moment create the quality of our experiences.

Fear and doubt do not exist on their own. We imagine them, and we use them to disempower our creative ability. Within our own consciousness, we have help with this condition. It is the Source of our vitality and is life-enhancing in every way. When we search for it, we become aware of it by aligning with its emotional and visionary stimulation. Intuition is the energy of the heart of our life-essence and our presence of awareness. It is our creative core and our connection with infinite consciousness.

Mastering our ability to create any reality for our experience depends upon our ability to express the vibrations that we want to experience. Even though we share the experience of the empirical spectrum of dualistic vibrations, we all imagine our own reality. We can change it at any time, within our personal destiny as we planned prior to incarnation. We can be aware of our destiny by intentionally penetrating into deeper memory. With our intuitive prompting and knowing, we have help fulfilling our intentions.

Once we feel that we've had enough experience of living in fear and stress, we can open our awareness to experiencing what we love. We can recognize, resolve and release our self-created limitations that are all based on fear of the unknown and possible conscious termination. As we go deeper into our conscious awareness, the unknown dissolves into infinite presence of awareness flowing through our heart-consciousness. Here we find fulfillment of all of our life-enhancing desires for all life.

The Nature of Our Divinity Within

If we want to recognize the true essence of who we are. We are the One who sees through our eyes and experiences everything we encounter. In order to have the intense experiences of physicality, we chose to impose limitations on our awareness, resulting in our being unaware of realms beyond the physical. If we have had experiences of them, we haven't believed they're real. These limiting beliefs are not required of us. They're just programs in our consciousness that are not needed any more, and can be changed.

We are experiencers having our own presence of awareness here and beyond time/space. We can express ourselves however we desire, except for our limiting beliefs about ourselves. These are programs that we designed to be unbreakable, but there is a key that allows for their recognition, acceptance, resolution and release with transcendence. It is found in our intuition, and it is awareness of our conscious life force and our connection in infinite consciousness through the energy of the heart of our Being.

The energetic quality of our inner guidance is the energy of our Creator, whose consciousness we receive and participate in. It is our inner-most knowing and our deepest feelings. It is the energy of creation and life-enhancement and all of the emotions that are present in this vibratory dimension. To be aware of it, we can imagine its reality in our alignment with gratitude, joy, compassion and love. This opens our awareness to experiences that we want to have. All we have to do is believe that these energies are true and realize their reality. When we shift our belief system, everything in our experience changes.

Limiting beliefs keep us locked into dualistic energetic patterns, which enclose our awareness, until we can realize our intuition and its value. It is part of our presence of awareness. When we follow our inner prompting and guidance, we can be free of limitations. We can live in a kind of parallel dimension based on the enhancement of all life. We can be constantly creating whatever we desire. This is what we do, whether we are aware of it or not. If we want to master this ability, we need to be able to direct our attention consistently to how we feel deep within. We can always know how to align our imagination and emotions with the vibrations we love and want to experience. Every scenario we want to experience already exists in the quantum field for us to recognize and make real in our experience.

With our energetic alignment, we attract resonating experiences. Everything we think about and feel has a vibratory quality. We have the constant opportunity to focus on whatever we want to feel and pay attention to. Everything is possible and available for our recognition through our inner guidance, which is always directing us toward creation and enhancement of life, including ours.

Dealing with Negative Situations

In our incarnation into this realm of duality, one important experience is encountering deeply negative energy and dealing with it. We naturally try to avoid it, because we deeply know that it diminishes life. We feel much more deeply emotionally than we would if we were only observers. With our participation, we have expanded universal consciousness into interaction with negative, life-destroying energy. Our interaction is what's important. We have the opportunity to be creatively life-enhancing and observant, to submit to it, or to engage with the negative by resisting it. This would bring our energetic resonance into alignment with the negative. It then takes our life force and may even enslave us. We know all of this on a very deep level, and we must learn how to interact with it successfully.

Because unconditional love is the essence of our lives, it is the most powerful force in our life. We are being challenged to realize this, which is our path for transcendence to a higher dimension of greater realization. We have the opportunity of experiencing the power of life-enhancing love in every moment.

If we meet someone who has been deeply hurt and, as a result, is deeply negative, assuming that this person has some ability to love, and if this person is insisting on engaging with us, perhaps threateningly, we have some options as to how to deal with the situation. In order to engage with negativity, we must do so in its vibratory spectrum. It is the nature of that band of energy to be destructive. Attempting to be positive at the level of ego-consciousness is doomed to failure. Another option would be to treat this person as a cherished sibling. When we expand our realization of what is real, we do not accede to negativity, and we know that love is the natural energy of this situation. Everything arises out of infinite love. The negative person cannot engage with this, unless (s)he rises to our vibratory level; otherwise, (s)he will disappear from our experience, because we are not energetically in resonance. The negative cannot reach us. It is a different polarity.

This outcome assumes that we are heart-centered always, and we express this energy in every aspect of our Being. By making this our predominant vibratory level, we align ourselves with the rising vibratory level of the Earth. When we can realize that we are the creators of our experiences, we can free ourselves from any kind of dependence beyond our own consciousness, and we can know the expansiveness of infinite awareness.

All the assistance that we need is available to us in realizing the essence of our Being. It is all within our own consciousness. We are each an eternal presence of awareness experiencing life as a human. We are the one who hears through our ears and receives sensory stimulation. We are our awareness of ourselves, able to play any roles that we desire for the creation and enhancement of all life.

The Next Evolution of Time/Space

As we enter the era of crystalline time, linear time, as we have known it, ceases to exist in its 3-dimensional form. We will still have time, but it's intimately connected with space in equal arrangements, creating a 4-dimensional experience. This is illustrated by time crystals, which physicists first imagined in 2012 and created five years later. Everything is changing in alignment with crystalline structure. The atoms and molecules that comprise our DNA are aligning themselves symmetrically for perfect functioning and expression of vitality. If we can think like quantum physicists or esoteric mathematicians, we can imagine a world of four dimensions. It is not linear. It is energetic and can manifest according to our openness to its energy and willingness to recognize it. It requires transcendence beyond our limiting beliefs about ourselves and is an evolutionary leap in our conscious awareness.

Now that time crystals are within the awareness of humanity, their energetic appearance is having an effect on human consciousness. Our presence of awareness is expanding into universal consciousness, within which we arise in our essence and receive infinite creative ability. Because we are quantum in our essence, we can realize our own awareness within expanding consciousness. Whatever experience we want, we can create by aligning our attention with its vibratory quality in gratitude and joy. Our creative ability is limited only by our limiting beliefs about ourselves. We can resolve and transcend them in a dimension without fear. Negativity cannot exist in the geometric structure of crystalline time, because negativity is an entropic anomaly in consciousness. There is only the life-enhancing energy of unconditional love.

Not knowing who we truly are allows for suffering, anger and fear, and we believe that everything ultimately ceases to exist. We are free to think and feel however we want, and to believe whatever we want. In doing so, we create the qualities of our experiences. There are no requirements of any kind imposed upon us outside of our own consciousness. We fill our lives with the energies that we pay attention to and align with emotionally, whether positively or negatively. What we are doing is creating new or repetitive experiences for ourselves. Our experiences on this planet have taught us that we prefer to be free of negativity, and instead to live in love and joy.

In our ego consciousness this kind of world is unimaginable, because the ego consists of limiting beliefs about ourselves. All of these must be resolved and transcended for us to evolve to the next level of awareness without fear. By paying attention to life-enhancing people, energies and scenarios as we move through our lives, we create the energetic environment for conscious expansion in infinite awareness.

Overcoming Survival-Consciousness

All of our limiting beliefs were created first in our imagination, until we believed that they are real. This is also how we can resolve them. They are all based in fear of the unknown beyond the physical world. From within the consciousness of the ego, we are not aware of what happens when our connection with the body is terminated; however, intuitively we know the extent of consciousness, because our intuition is multi-dimensional, as are we, once we realize it.

Everything can be real. The unified quantum field of conscious expressions, within which we exist, has every possibility of everything. Whatever vibrations we recognize and pay attention to are our reality. Imaginary or experiential are the same vibratory patterns. What's important is what is in our attention and how we feel about it. We can create whatever beliefs we want in our imagination, as well as resolve those that we've acquired. If we do this powerfully and long enough, we realize their reality in our experience. By clearly examining the limiting beliefs that keep us from realizing our true Self-Awareness, we can find their basis. If we choose to replace fear of the unknown with a change in perspective to awareness of the life-enhancing creative energy that envelopes us throughout infinite consciousness, the unknown becomes known and fills us with gratitude and joy. The limiting beliefs dissolve.

Our awareness is limited only by our own choices, and our experiences come to us as a result of our realization of what is real. What is important is not the subject matter of our thoughts and emotions, but the vibratory quality, the polarity and frequency. We express our state of being by choosing depression, uninspired existence or vitality. If we choose vitality, we feel and radiate gratitude, love and joy in every moment, and we live with abundance and freedom in infinite presence of awareness, realizing support and knowing everything we need.

As we learn to be aware of the vibrations and prompting of the heart of our Being, everything becomes easy, and we can live without stress. We can transform our lives into a higher energetic dimension without going anywhere physically, but the quality of our lives changes dramatically. We can live in life-enhancing vibrations that bring fulfillment, and we can share our vitality in our encounters, even the challenging ones.

We are never required to align with negativity, and once we are living in gratitude, compassion and love in every moment, negativity cannot come into our experience. We are sovereign beings arising from an essence beyond time and space, having unlimited creative power through the consciousness of our heart.

Realizing the Greatness of Who We Are

Although our awareness is potentially infinite, in our lives as humans, our awareness is only within the limits of our beliefs about ourselves. If we can imagine what part of us lives beyond our physical presence, eventually we can realize that we are our presence of awareness, and we are unlimited in every way. At first we don't necessarily know our abilities, or even the presence of our inner knowing. For eons we've been living haphazardly under the guidance of the ego. We can transcend ego-consciousness by focusing on the energy of our heart. It is our conscious life-source, empowering us to live the way we believe we should within any belief structure that we have accepted as true. It is our choice whether to believe that we are rich or poor, free or enslaved, normal or handicapped, bright or dull. Powerful beliefs can be thoroughly examined down into our subconscious to find their origin. This is where we hold our deepest fears, which anchor our lives to negative experiences and limitations. Fears are based on a misunderstanding of reality and ignorance of our essence.

If we wish to know these things, we must open ourselves to their energetic radiance. It immediately becomes clear if we're dealing with positive or negative energy. If it's positive, we can trust it to be life-enhancing, stimulating us to feel good in its presence. In this state of being, we can use our power of choice to imagine that we are living in the energetic expression of our heart and feel its vibrations. We can do this in every situation in the best way that we can. Eventually we can be confident that we know how to be and what to do always, as we practice following what we feel in our life-source.

In our thoughts and emotions, we can choose to be grateful and life-enhancing for all within our energetic radiance, and we can express ourselves in alignment with our intuition. By living this way, we can transform our lives from the situation we have become accustomed to, to a life of fulfillment and joy.

When we realize that our empirical world is an expression of consciousness that we have aligned with energetically, we can begin to live with intention and freedom. We experience living in the qualities of our mental and emotional processes. It all happens within our own realization, by believing what we believe. By recognizing, resolving, and transcending our negative beliefs about ourselves, we open ourselves to realization of our infinite awareness and creative ability

Existing forever in universal consciousness, we are free to realize as much of it as we choose. The current of our life-force is carrying us to renewal, regeneration and Self-Realization as fractals of infinite creative, life-enhancing consciousness. It is the same life-stream shared with our planet and all the beings beyond. We are being carried into a new era of love, beauty, joy and fulfillment in every way.

Exploring Our Belief in Good and Evil

In order to understand our own reality, it may be helpful to know what quantum physicists know about the structure of the empirical world. First, there is no such thing as substance. Everything that we are aware of, at its source of being, consists of conscious, swirling patterns of energy, both etheric and electro-magnetic. This applies to the atoms of our body and all subatomic entities. They are swirling energetic expressions of their own conscious awareness in alignment with the consciousness of our body, which is an expression of the consciousness that creates everything.

Although we are not our bodies, we are intimately conjoined, so that our body expresses the vibratory patterns of our accustomed state of being. We have believed that the flesh of our body is solid, but penetrating its structure down to its smallest constituent parts, reveals a complex energetic expression of trillions of conscious subatomic entities of light-essence, some of which vibrate in frequencies within our sensory range. We can recognize them in great numbers of swirling patterns of energy moving at light-speed, appearing and feeling solid to us.

At the source of everything is consciousness. Everything is conscious within its own structure. Consciousness is everywhere, and we are part of it in our own ways and to the extent that we allow ourselves. We have ignored the consciousness of nearly everything, including the constituent entities of our own body. We can communicate with them, and we can influence their vitality with our own perspective. Whenever we express life-enhancing thoughts and feelings, the cells and DNA of our bodies naturally cooperate with us. When we are negatively-oriented, we radiate life-diminishing energies throughout out body.

In the empirical world, in order to have matter, we must have anti-matter as a balancing backdrop, or there would be no awareness of matter. That allows for duality, for creation and destruction. These are both necessary to contain the expression of infinite Being within the confines of the dualistic empirical band-width and polarities. In order to participate in a convincing way in this world, we have imposed upon ourselves the belief in good and evil, light and dark and all polarities, but we are not required to maintain this perspective.

We know that our vitality is creative and life-enhancing. We know when energies are destructive and life-diminishing. This is part of the human experience in duality. If we want to open our awareness to greater consciousness beyond duality, we must align our attention with the energetic patterns of the source of our vitality. We can intentionally seek this awareness without attachment to any beliefs, opening our inner awareness for what we truly know as life-enhancing gratitude, love and joy. These are the qualities that arise within us in the infinite consciousness of the Creator. They are the vibrations that open us to universal consciousness through our intuitive ability.

Creating a Life of Beauty and Majesty

In our essence, we have always lived in beauty and majesty. In this moment we live in absolute power over our lives in infinite awareness and infinite love and joy. In order to be aware of our true Being, we must arise from our human hypnotic trance into a state of being that is terminally feared by our ego-consciousness, because it is unknown. To our ego-conscious mind, there is great danger, if we release the control of our beliefs about ourselves. The fear of termination of ego-consciousness is real, because the ego-mind is based on fear of separation from the Source of our life. Without that fear, there is no ego. There is only our intuitive Self, our eternal presence of awareness, able to create everything our hearts desire, including our human persons and everything about us.

Creation of our personal expression and our life experiences in every way is what we do, whether we realize it or not. If we want to live in beauty and majesty, we must align our vibrations with their energetics. We can use our imagination and emotions to carry ourselves into scenarios with wonderful, loving encounters. We can use our imagination to realize the Creator-light in everyone, even the evil ones. This presence of Creator consciousness is in everyone. Anyone or anything beyond our own essence has no power over us without our permission. Even so, we cannot be destroyed, because we arise in every moment, trillions of times each second, within the consciousness of the Creator. We share in this consciousness, which is everywhere and in everyone. We are all conscious together and can be aware of our own and each other's awareness.

The energetics of life are constantly creative, and any kind of fear that blocks our life-force is an anomaly and can be resolved in alignment with intuitive knowing of our Self. Everything in our awareness has a vibration. When we focus on one, we align our vibrations with it. They radiate out into the quantum field with instructions for the kind of energy we want to experience. When we are negative or positive toward others, we are sending that energy to ourselves, resulting in our experiences.

There is also the matter of destiny. Prior to our incarnation we decided what kind of experiences would most benefit us to awaken to our true Self, given that we have no memory of who we are. In this sense, our lives are symbolic metaphors that guide us toward this goal, constantly offering and urging us into awareness of the consciousness of our heart and the vibrations of our life-force.

Aligning with our deepest intuitive knowing connects us with awareness of our infinite essence, having every possible ability and constantly using all of them according to our vibratory state. Our ability to create experiences for ourselves requires only that we vibrate at a predominant wave-band, determined by our mental and emotional processes. By imagining and feeling that we are in scenarios expressing gratitude, love and joy, we can align with our intuition, in which everything is known.

Our Deepest Love

We're learning how to be loved, truly loved, because we are. It's time for us to awaken to our reality. Love and joy are our nature. We do not need to receive them. We already have them in their fullness, because they are the essence in every cell and atom of our physical presence and our entire consciousness. We only need to recognize them and be aware of them. They are most radiant in the consciousness of our heart. We know them through our intuition as the most life-enhancing thoughts and feelings that we have.

We are creators and energetic modulators by the vibrations of how we feel about ourselves in every moment. Everything we could ever want is available in the plasma field of energetic patterns that envelopes us. When we align our attention and emotions with the vibratory spectrum that we deeply desire, our fulfillment comes into our experience. If we want to experience deeper love, we can look into our essence, to our deepest intuition in the energy of the heart and source of our conscious life force, our inner knowing. It is the consciousness of our heart, physical and etheric. It is eternal, infinite love and joy in our presence of awareness.

Every experience in our lives originates in the vibratory spectrum of our state of being, consisting of how negative or positive we are in our thoughts and feelings about everyone and everything. Our emotions are magnetic, and they attract their own quality of vibrations. When we fill ourselves with gratitude and love in every moment, regardless of the energy around us, we bring their energies into our experience. By being grateful for every experience, and by loving everyone and everything, we remove ourselves from subjection to life-diminishing experiences.

We live enveloped in unconditional love, providing everything we could want. By aligning our thoughts, beliefs and emotions with everything that enhances life, we participate in the creative consciousness that is the source of our life. By resolving and transcending our limiting beliefs about ourselves, we gain access to infinite awareness and the unlimited creative power of our true essence. We can be the directors of our lives in complete sovereignty and freedom, expressing goodness throughout our entire Being. We can choose to expand our awareness beyond our deeply-set limiting beliefs and be in alignment with the infinite consciousness that we participate in, and whose vitality we constantly receive through the consciousness of our heart.

Examining Our Core Consciousness

If we want to be aware of everything we need and want to know in every moment, we must align ourselves with our intuition. This is where we absolutely know what we know. There is no outside proof of our reality. Although we can create the proof, if we want to, it is all within our own consciousness. We have done this with the empirical world. We have given it reality for us by our realization. Now our life stream is carrying us beyond our limitations and beyond time/space as we have known it. We are being invited into a realm beyond polarity, where there is only joyful and loving creativity. Intuitively, we can direct our thoughts and feelings in alignment with life-enhancing energies in every moment.

Knowing our essence beyond the limits of empirical duality, we are no longer subject to those limitations. We are pure eternal, conscious presence of infinite awareness and creative power. Within the world that we have realized as real, we have enclosed our awareness. We can be within this awareness, while at the same time be aware of our infinite essence of Self-Awareness, knowing everything about what we pay attention to, as well as the entire consciousness of humanity, our species consciousness. Beyond space/time, we are all our own presence of awareness, the source of our conscious life-force.

We evolve moment to moment in our conscious realization.

We are either increasing our vitality and radiance of gratitude and joy, or we are diminishing ourselves. There are no requirements outside of our own choice and desire of how we want to feel in relation to others and with ourselves. When we know who we are, we do not have to feel bad if someone belittles us. Such encounters are inviting us to transform the negative, life-diminishing energies into enlivening energies. It happens within our own consciousness by our intent. All possible energies and situations are potentially real in our experience by our recognition and realization.

Although we can create forms with our visions, the qualities of our lives depend upon our predominant vibratory state expressed within our consciousness. When we live in gratitude, joy, compassion, love and other life-enhancing energies, we attract wonderful experiences and relationships. We can resolve and release all fear and limitation and be confidently open to our intuition in every moment. Intuition arises in the conscious life-force that we receive through our heart, and we direct its use at the vibratory level that we choose to pay attention to.

Living in the empirical world can have much enjoyment through our senses. We choose how we want to experience it by our vibratory level. We control our perspective and our beliefs about ourselves. Our perspective can always be based in loving creativity. Once we can realize all of these things, we are no longer limited in any way. Our conscious awareness transforms without stress or fear into a realm of beauty and magnificence, dimensionally beyond negativity.

Accepting Our Invitation to Conscious Expansion

When we're in a beautiful and serene place in nature, we may want to drift off into the eternal One, connecting to the awareness of all conscious beings around us. Intention and practice enable this awareness in us. While we are all the same consciousness, we each have our unique personal awareness. If we are curious about what may be beyond our personal awareness, and we get serious about it, and we search for it and are open to know and understand, guidance will come as prompts from our inner knowing and from our continuing circumstances. We always have guidance for finding our way to realizing who we really are in the depth and expansiveness of our consciousness, and the quality of our true essence.

When we imagine living in gratitude, love and joy, we create experiences that elicit the same feelings. When we imagine our awareness being drawn into alignment with infinite consciousness, our emotions become vast, and we become aware of our creative power. In this state, we can transcend our limiting beliefs about ourselves, because they are not true. All of them have the essence of fear, which we can have only if we create it for ourselves. It has no essence of its own, apart from the lifeforce that we give it through our attention and alignment. Without our intention, it could not exist for us.

When we stop paying attention to the compromised media,

we help to dissolve the media's negative energy. It is presented as an appearance of truth, but it's really mostly lies and anomalies of the truth. Even the movies and TV shows are designed to contain our awareness. The media cannot exist without our attention. By withdrawing our attention, the negative energies of all kinds drift away from our experiences. In our experiences, our thoughts and emotions can always be open, curious and observant of our inner guidance, regardless of what may be happening outside of our immediate situation.

When we decide that we want to expand the radiance of our heart, we can open ourselves to our deepest knowing and feeling about everything. As we become proficient in our intuitive awareness and guidance, life becomes fun and fulfilling. We can realize our eternal presence of awareness constantly arising in us. By paying attention to this awareness, we can keep expanding into new and enlightening experiences.

Transcending our self-imposed limitations of every kind allows us to realize our divine essence and infinite creative power. We can realize our awareness beyond space/time, and we can be always compassionate, loving and joyful. This level of vibratory radiance attracts its reflection from the quantum field of all potentialities and provides a dimension of living with only life-enhancing energies for us to enjoy while strengthening the heart-consciousness of humanity in alignment with the consciousness of the Spirit of the Earth.

Insights into Universal Consciousness

Our ego is our programmed consciousness, which contains our awareness. In our ego-consciousness, we do not know what is beyond time and space, because the ego does not recognize our inner knowing. This allows us to feel separate and be fearful of the unknown, particularly what happens after the body dies. This fear is held in our consciousness and pervades our awareness, keeping us believing that we are insignificant, shameful, victimized and terminal. It is all in our imagination and can be changed, if we desire to evolve. When we recognize the power of our beliefs, we can be grateful for the gift of this power, and we can learn to direct it in life-enhancing ways.

To free our ability to direct our lives in confidence, we can examine the basis of our limiting beliefs about ourselves. They hold our awareness locked into negativity. We hold ourselves in fear and doubt about our lives. We do not know what may happen to threaten us. This anxiety keeps us from realizing our higher guidance within and everywhere. When we feel threatened, we cannot have an open heart, even though the threat must be self-created, because we are eternal, sovereign Beings. We have the ability to transform these limiting beliefs by realizing the unconditional love vibration at the heart of our Being, symbolized and embodied by our physical heart, which lives to enliven us, regardless of what we do to it. The heart has been shown medically as having by far the strongest energetic radiance of any organ in our body, including the brain. The heart is the source of our intuitive knowing and is our connection with infinite consciousness. Intuitively we can know whatever we want to know in every moment.

Consciousness is infinite and is in everything. Trillions of times each second, consciousness creates every form and substance that exists, including our essence. It enables our personal expression through the beliefs that we imagine and apply to ourselves. Consciousness lives beyond all energetic expressions and is within them. We may think of universal consciousness as a living Being that we arise within and are given freedom of thought, emotion and attention. To us, this Being may be God the Creator or Allah or the Void within which all arises. In the realm of science, quantum physicists began to recognize universal consciousness in the early 20th century.

Although each of us has our own awareness, we all participate in infinite consciousness, including everything manifest and unmanifested. To open our awareness to infinite consciousness, we must resolve, release and transcend our limiting beliefs about ourselves. We needed them for our human experience, but when we are ready to expand beyond the world of good and evil, we can recognize that all of life, including the Earth, is trending toward compassion, love and understanding. If we are stuck in negativity, we will have to search intentionally for positive and higher vibrations to align with. When we align mentally and emotionally with the energies of gratitude, joy and love, we are enabled to see and feel the light in all beings and to connect with their awareness.

Shifting beyond Ego-Mind into the Unknown Void

In every moment we walk with the divine. We suffer because we have been socially programmed to suffer, and we do it because we haven't realized that we're doing it to ourselves by our beliefs about ourselves. Everything begins in our imagination and continues into our experience. If we choose to take the initiative with our intentions, we can transform our limiting beliefs. In our essence we are capable of extraordinary experiences. Although we are unlimited in our capabilities, we have not believed this. In order to master our lives, we must be able to believe in the reality of what we want to bring into our experience, and we cannot rely on our ego-mind for this. We must expand into our heart-consciousness.

To be attuned to our heart-mind, we must learn to use our realization differently from the way our ego-mind operates. In the heart-mind, there is no calculation or analysis, there is only instantaneous knowing. It is beyond even using both hemispheres of our brain. By activating our full cerebrum, we can bring our ego-mind to open to our heart-consciousness by intentionally using our imagination to bring us into a state of gratitude for everything. We can open ourselves in confidence to recognizing our guidance, which is always life-enhancing and absolute.

All vibratory levels arise within our awareness and imagination. Once we can feel the vibrations of our heart's intuition, we can shift beyond fear, intimidation and stress. We can realize that we are the creators of our personal experiences and of our world. Realize it or not, we control our personal vibrations, which attract our experiences. Our vibratory limits are set by our perspective, our beliefs, and how we feel about ourselves.

Any vibrations created with negativity are self-destructive. They block the radiance of our heart and limit our conscious life-force. In order to expand our awareness, we must change how we use our power of attention, directing our awareness to energies that are life-enhancing, with the intention of being aware of the divine life-force in every encounter. This is the transformative power that elevates our lives and our relationships. Our personal dramas disappear. Replacing them is personal fulfillment of all needs. We can enter the realization of infinite consciousness and the awareness of all other awarenesses.

With our ability to feel and read the qualities of energies facing us, we can accept it all in gratitude and joy, for we have at some level wanted to experience everything, and so we created their energetic patterns. They are illustrating for us our creative power, even when it's unintentional. By recognizing our inner guidance and aligning our vibratory level with its energetic expressions, we can shift into greater awareness of our connection with infinite consciousness.

Choosing and Experiencing Our Quality of Life

I met a fascinating person today. He was a beggar of sorts, but he had no begging bowl, and he stood tall and Self-assured. He asked for nothing, but his presence spoke. I knew immediately who he was by the light in his eyes. He chose to stand in front of the local health food store, as beggars sometimes do. Some of them play music, some entertain, and some are just desperate. He was just himself. He wanted to meet heart-conscious people who are fearless, and who recognize him and are moved to converse. He wanted to affect every person who walks by with the psychic challenge for them to realize their state of being.

This is our challenge as well. What kind of energy do we express in every moment? Are we aware of it? Taking a step toward enlightenment includes accepting everything about ourselves and realizing the fears that created our self-limiting beliefs. We have allowed ourselves to create situations of fear and intimidation. Upon examination, we can understand that they have no basis outside of our own consciousness. Through our attention and emotional vibratory alignment we have allowed fear and negativity to enter our state of being, closing off awareness of our heart-consciousness and inviting negativity into our experience.

Without heart-consciousness, we are subject to ego-consciousness, which reasons and ponders based on its history, memories and logic, but does not deeply know anything. It is constantly under stress and fear of the unknown and of termination. We do not have to subject ourselves to poverty, fear and feelings of separation from others. These conditions all stem from our limiting beliefs about ourselves, and can be resolved by intentional awareness and our desire to know and feel our heart-consciousness through our intuition.

We have been given the choice of using our creative essence in whatever way we want and imagine, just by the polarity and vibratory presence of our thoughts and emotions. In every moment we are creating our experiences by the way we are, mentally and emotionally. Our experiences have taught us to know all about the energetic patterns of the dualistic, empirical world. We can recognize them all, and we know which ones we want to experience. By choosing to live intentionally in the energies that we love, we no longer experience those that we do not dwell upon. Even though they may be happening for others around us, they disappear from our lives.

We Are Energetic Transformers

For eons humans have lived without knowing our true nature and capabilities. We have believed that we are separate individuals with our own private consciousness, limited abilities, and dependence upon society for our nourishment and well-being. We have believed that we are physically subject to threats, suffering, intimidation and termination, not knowing that all of this is created in our own perspective and belief system. We've had numerous spiritual masters who taught us that our fears are all imaginary. Now we also have quantum physicists, who have reported that consciousness is infinite, and that the physical world is actually an energetic expression of consciousness. The empirical world is actually composed of swirling subatomic entities of living light. Everything is conscious and participates in universal consciousness according to its self-identity.

We are no different. We participate in universal consciousness according to our Self-Realization, which can be infinite, but has been limited for eons by our beliefs about ourselves. These beliefs are deeply-set in our personal awareness and can be nearly impossible to transcend, unless we are powerfully motivated to know our true essence. Now the rising cosmic vibratory environment is prompting all of us to become aware of our potential. Our accustomed way of life is being destabilized and is dissolving and transforming into a higher dimension of increasing light and beauty. Although it is still possible for us to choose to remain in the grip of our limiting beliefs about ourselves, our planet and all who choose to ascend are in the process of transformation. It is being accelerated by increasingly powerful gamma ray photons and neutrinos from the center of our galaxy. As the constituent conscious subatomic entities comprising our bodies gain vitality, our cells and organs begin to regenerate.

All of this may be happening without our conscious awareness, but we can contribute to the magnitude of this transformation by opening our awareness to the intuitive knowing within the consciousness of our heart through our intuition. By directing our awareness to what we deeply know without any outward proof, we can imagine living in gratitude, love and joy. These vibratory patterns align us with the life-enhancing energetic expression of our heart-consciousness and guide us beyond our limiting beliefs.

At some point, we can realize that consciousness is expressing itself energetically everywhere and in everything, and that our ability to recognize the energetic patterns brings them into our thoughts and emotions. By using our imagination and feelings intentionally, we can transform negativity into positivity or dissolve it by holding our attention on the intuitive guidance of our heart-consciousness. By our nature we are the transformers. Whether we realize it or not, we are constantly creating the qualities of our lives and contributing to the consciousness of humanity and beyond. Now we are being invited to do this intentionally.

Directing the Living Play of Energies

The nature of dark energy is to absorb light and life. Dark energy is the source of anti-matter and everything that is the opposite of matter, light and vitality. We would not know about light, if we did not know about the dark. For any kind of expression, there must be an expression of the opposite polarity, because the universe is striving for balance and adjusting to anomalies. Humanity has been an anomaly of dark energy that threatens to self-destruct by becoming an energetic black hole that absorbs the light and life-force of our accustomed reality.

As we realize our greater reality of light and vitality, we can leave the dark, negative frequencies to be unexpressed in our presence. They still exist in potential in the quantum field, but they are not in our experience. We can choose to have constant awareness of what we deeply know and feel about our own nature. We can choose to live in alignment with gratitude, love, compassion and joy. These are all part of the nature of the light of our life-force, which we receive constantly from the consciousness that we arise within.

It is possible to know ourselves, our capabilities and the vastness of our non-localized presence of awareness. Our true nature is beyond the comprehension of our ego-mind and our realization of reality, but we can open our awareness to our essence through our intuition and heart-consciousness, by aligning ourselves with life-enhancing expressions. These are the vibrations of our true nature, and we all know how they feel and how desirable they are. We are naturally drawn to them, and we can recognize them clearly because of our exposure to darkness and negativity. Through this realization, we can begin to trust ourselves to use our creative ability for life-enhancement everywhere.

When we realize our true nature and the essence of our consciousness, we can resolve and release our limiting beliefs about ourselves, because they no longer apply to anything. What is real for us is what we realize is real. As long as we live in the empirical matrix, it provides the stage upon which we play our roles in the human drama. The qualities of our lives result from how we feel about ourselves and the energies that we project with our thoughts and emotions.

By paying attention to the life-enhancing expressions of our heart-consciousness, we can realize that we are free, appreciative and fulfilled in every way. Our intuitive knowing guides us in every moment to understand our situation and receive what we need and desire.

Intentionally Changing Dimensions

Consciousness is a non-localized presence of awareness with memory, intentions, interests, understanding and creative ability with mental and emotional processes. Being non-localized, consciousness provides potentially infinite awareness. Our essence and all that exists for us occurs within infinite consciousness. We may infer that infinite consciousness is part of a greater Being, whom we can know through its energetic expressions, which come to us through the heart of our Being, and through the expressions of our pets and all creatures whom we are intimate with. The same can be understood for the plants and trees that we live around and interact with on subtle levels.

We have been accustomed to keeping our awareness enclosed within an energetic box of beliefs about who we are. This has been necessary for our experience in the empirical world of good and evil. Without the experience of darkness and death, we did not know what they are and what their relation to light and love means. Now we have a deeper understanding of life and the importance of wanting always to expand love and joy in the enhancement of all life. We no longer need to subject ourselves to negativity. Our participation has always been voluntary in the depth of our consciousness. When we become aware of this, we gain the ability to choose how we want to use our awareness and how we want to create our lives. We can realize that we can be the absolute directors of our energetic participation in every experience and in our imaginary visions and feelings.

Instead of seeking pleasure and avoiding pain, while struggling to survive, we can change our lives to feeling fulfilled in every way and being grateful and loving in every encounter, and in our random thoughts and feelings. We can come to know our true essence as participants in universal consciousness, unlimited in every way and able to fulfil everything we need and desire just by imagining having them in our presence and feeling ourselves experiencing them with gratitude and joy, like when we return home and our dog jumps for joy just for our presence.

In the dimension of life-enhancement, there is no fear or any negativity. This is the realm we are being drawn into. It exists right here and now in a higher vibratory spectrum than the enclosure of consciousness that humanity has lived within. By intentionally living in the conscious expressions of our heart, we can experience the magic and joy of wonderful experiences always. By creating a vibratory signature based in gratitude, joy, compassion and love, we attract those energetic patterns into our encounters, bringing fulfillment and greatness into our lives.

During our transition from fear to love, we remain aware of all the negativity happening among humanity, but as we focus our attention on creation and enhancement of all life, negativity and negative people magically disappear out of our lives. It's not really magic, but a change in energetic octaves. Making this change requires intuitive knowing that it is true and realizing it in our experience. What we realize as real is an expression of our choice of what we believe.

Insights into Our Expansive Potential

At some point we realize that we have a presence beyond the physical world. We have imagination and emotional feelings. They are not abilities of the body; rather, they direct the processes of the body. They are related to our essence and connected to our expanded consciousness. They are part of our presence of awareness, our creative center. It is where we shape the qualities of our experiences and the intensities of their manifestation. This happens in our deep consciousness, guided by the vibratory quality of our beliefs about ourselves. When we have this awareness, we can clearly understand how we create every nuance of our experience.

As we bring into our attention the thoughts and feelings that we choose in every moment, we can intentionally align ourselves with these energies and live in their range of vibrations. If we want to expand our awareness beyond our acquired limitations, we can use our imagination and emotions in the service of our heart-consciousness. This is the source of our conscious life-force, and it is completely life-enhancing in creative ways. When we open our awareness to the influence of heart-consciousness, our desire draws it to our attention and allows it to become the dominant energetic quality in our awareness.

Since we have limited our reality to our sensory experiences and our encounters with others who have the same limiting beliefs about themselves as we do, we have lived entirely within the compartment of consciousness circumscribed by our beliefs, with no awareness of a way out. Our only guidance is our mental calculations and our awareness of history, together with an innate desire for transcendence. If we pay attention to this desire, we can open our awareness to our radiant essence and presence of conscious life-force.

When we pay attention to how we feel about anything, we immediately know if we are facing negative or positive energy. By keeping our attention on our transcendent essence, we can feel the warmth and enrichment of our heart-consciousness in universal love, compassion and joy. When we feel these energies, we are open and aligned with the intuition of our heart-consciousness, and everything becomes clear, understood and fulfilling. This is how our inner knowing develops.

Processing Our Limitations

Our beliefs about ourselves are the primary confinements of our consciousness. Our awareness does not go beyond them. They define how we are in every way, right down to the vitality of our subatomic physical constituents. When we align ourselves with life-enhancing energies, we do not need limitations. We go through a process of transcendence, finally realizing that we are infinite in our presence of awareness. We envelop the consciousness that constantly creates universes. We share in infinite creative power.

While we have been experiencing negative, life-diminishing thoughts and emotions, we could not trust ourselves with our awesome abilities. We hid our true essence from ourselves in our heart's consciousness. The heart has intelligence far beyond the capacity of the brain. It does not need to think and analyze, because it already knows everything. Except for our limiting beliefs about ourselves, we already know everything, and we can feel the awareness of every being around us.

We are in a process of expanding our awareness. We are awakening from the hypnotic trance of humanity. Through our ability to call forth thoughts and emotions, we can control our state of being. This is our energetic signature. It constitutes the expression that we radiate within and around us and attracts compatible patterns of energy, which become our experiences. When we pay attention to thoughts and emotions that turn us on, we align with these energies and attract experiences that are formed by a similar expression that is in resonance with us. In this way we can transcend our fixation on living within limitations.

Instead of believing that we must struggle to survive, we can open our awareness to the consciousness of our heart and follow our intuitive guidance. It will take us into experiences that are fulfilling and wonderful. We have to work through our doubts about our creative ability. Doubt itself has no reality of its own. It is our own personal creation, along with all forms of fear. It arises from our belief in mortality and our subjection to negative energy. This is a choice on our part, and we can change it. We can realize that it is impossible for our awareness to end, because it is beyond time and space. Our awareness is always present in infinite consciousness, of which we are fractals, sharing in all aspects of its expansiveness.

Since our ego-consciousness consists of all of our fears and limiting beliefs, it cannot imagine that we could consist of infinite conscious awareness in the perpetual present moment, which includes everything. Our essence cannot really even be described, because words are limiting. As our realization opens though our heart-consciousness, we can understand how our enslavement to negative experiences became possible, and how we can allow them to dissolve from our reality, by paying attention to what we deeply desire and love in everyone and everything with gratitude and joy. This is the path of miracles and wonders.

Understanding and Expanding Our Humanity

Our beliefs are incredibly powerful, and they are the governing factors of our lives. Even though they are seated deeply in our subconscious, they can be entirely under the control of our conscious choice. We may have to use our imagination at first to get ourselves into a calm and loving space. Unless we know our true essence, we cannot master the human dimension. Our essence is our potentially infinite presence of awareness.

Our consciousness exists beyond time and space. We are the real masters of time and space, but only if we allow ourselves to be. If we can trust ourselves absolutely to be who we truly want to be, we can know our infinite intuition, and we can transcend the entire human situation in favor of a greater realization of our access to the quantum field of all potentialities. This gives us the ability to modulate the energy that creates time and space. This ability must come into our awareness, and we must direct it with our attention and emotions in alignment with our true intuitive knowing. This is how we come to trust ourselves.

All beliefs are false, because they are all limiting our greater realization and expression. From a limited perspective, we cannot allow awareness of our non-localized, infinite conscious essence. It is beyond our limiting beliefs about ourselves and our abilities. While we are beginning to realize the truth of our eternal expansiveness, we can still be subject to doubt about our creative ability. Doubt makes us vulnerable to incursions of negative energy into our experience, because it aligns our polarity with negativity, disabling our life-enhancing creativity, while creating life-diminishing energetics.

Doubt is our creation. In order to have doubt, we must have fear and the belief that we can be terminated. Beliefs create situations that confirm their existence. This is our challenge, and we will continue to have doubt and fear until we can trust that we know our eternal presence of awareness with infinite creative power in alignment with the vibratory patterns of our heart-consciousness. We are the creators of the qualities of our experiences, and our reactions create further experiences.

At some point, we can realize that human life is a play of consciousness with meaningful experiences. These are designed to show us our own energetic expressions and how they feel to those around us. Because human life is a play of consciousness, we can learn to direct our experiences in ways that create and enhance all life in alignment with the consciousness that creates everything, and which we are part of. Once we realize this, time and space become our medium of expression.

Acknowledging Our Inner Light

In our lives, light is provided by our Sun, which is so bright that we cannot look at it. Each of us is capable of this kind of radiance. Light is our essence, along with all energies that create life and make it better and more beautiful. All of these vibratory energetic patterns consist of conscious subatomic beings, expressing potential forms that we can modulate into experiences by our mental and emotional processes. We can see our inner light through our pineal energy center, and we can realize it in the vibratory qualities of our heart. This is where our conscious lifeforce arises in the consciousness that creates everything, and we are the directors of its manifestation in the empirical world.

We have beliefs and perspectives that align us with resonant energetics in the quantum field. This is where manifestation is created, resulting in our experiences. Because we have subjected ourselves to a world with negative, destructive energies, we have veiled our inner light. Except for our beliefs, we are not required to maintain this veil. When we realize our non-localized essence in our presence of awareness, we can free ourselves to feel, know and express the unconditional love that flows through our heart. These are the energies that we are naturally most attracted to.

Although all beliefs are limiting, we can use them to achieve transcendence beyond them. By choosing to realize small miracles in our lives, we can set our beliefs a little further into our fulfillment. We can do things like align in gratitude with the angels of the air and believe that they love to cooperate with us, bringing about weather patterns that we desire. We can thank our personal angels for creating a parking space that we want and believe that it is happening. If we stretch our beliefs a little, and we continue to do so, we can eventually come to realize the infinite, creative, inner radiance that flows into us and that we can direct. We can cooperate with all the forces and beings of nature.

We set our own limits in everything, from the conditions of our bodies to the qualities of our lives and relationships. Much of this we have inherited, in order to have specific experiences, but once we come to realize our true essence and choose to align with its vibratory level, which we know through our heart-consciousness, we are free to create whatever experiences and life situations we desire. The radiance of our inner light becomes visible beyond our physical presence.

Empowering Our Transformation

We are being invited to realize what is beyond our ability to conceptualize, write or speak about. It is our own true essence, our Self-Realization. We have clues of our greatness in the masterworks of art, music, poetry and dance, as well as in the presence of mystics and spiritual masters. We all participate in this level of consciousness and so much more. Hidden in the consciousness of our heart, our Self-Realization transforms our lives and provides everything we could ever want. In realizing our own essence, we come to know the secrets of time and space and of infinite awareness.

For eons we have depended upon our conscious mind, our reasoning and our physical abilities to survive and evolve in a world of limitation. We have ignored what we deeply know about our reality and have put all of our attention on appearances. We have not known the source of our life and what our capabilities are. Because we have allowed others to train us in servitude, we have not realized that we are actually the creators of our conditions and experiences.

At this time in our planetary history, we are not here to experience sadness and suffering. We are here for transformation and expansion of light everywhere, and we are all capable of achieving this, or we would not be here learning about awakening to our true essence and wielding our infinite creative power. The enforcers of our enslavement now have power over us only if we cooperate with them and give them our life force through our fear.

Whenever we desire, we can withdraw our alignment from the dark force by focusing on the inner light and infinite love flowing to us through the consciousness of our heart. We can learn to recognize this energetic expression in everyone and everything. Everything that exists participates in the creative, life-enhancing consciousness of the Creator. This vitality is now the dominant energetic expression on our planet.

All of humanity must come into alignment with it in order to continue to live here. Those who choose to align with the dark force are becoming more uncomfortable and unstable as we continue to move further into the light of love and joy of expanding consciousness. We are moving through a period of desperation on the part of the negative elite, with all of their dissimulation, trickery and destructive intentions becoming known and eliminated. As we can align ourselves with the expressions of our heart-consciousness, we disempower any outer control of our lives and transform ourselves into the powerful Self-Realized masters that we are in our essence.

More Fully Developing Our Capabilities

By recognizing a reality, imaginary or experiential, we can align with its vibratory range and invite its qualities it into our lives, filling us with experiences of resonating vibrations. Through our ability to modulate energies, we direct the qualities of our experiences by the polarities and frequencies of our thoughts and emotions, which create our present state of being. How we experience life is entirely our own choosing. We choose our orientation and direction in every moment.

We choose the forms and expansiveness of our lives. Except for our vibratory reactions to outside stimulation, we can with gratitude, accept and love everyone and everything that comes into our recognition of reality, regardless of how our ego-consciousness may regard them. This keeps us in a life-enhancing realm of realization, in which we transform any negative energy in our lives.

The qualities of all of our experiences result from what we think about and feel every day, not the forms, but the vibrations. If we want to experience greater love and joy in our lives, we must realize their presence in our thoughts and emotions. They already exist in our conscious life-force, waiting for us to open our realization to their reality in our lives. When we intentionally seek awareness of the source of our life, we are guided to its realization within our own essence. It is in the depth of our consciousness, beyond thought, in our deepest knowing and feeling. It can only be realized in its fulness by aligning with its vibrations of life-enhancing creativity.

To live in the new world, we don't need to go anywhere or change anything. We only need to transform our realization. This changes everything in our lives that has had a negative vibration. By gratefully aligning ourselves with love and joy, regardless of the presence of any other energies, we can keep our connection with our deepest consciousness. As we gradually expand our awareness beyond time and space, and into our eternal presence of infinite awareness, we gain a sense of mastery of our lives in absolute confidence.

Only with our recognition, can something exist for us. By recognizing and aligning with specific energy patterns of scenarios in our imagination or experience, we interact with the quantum field of all potentialities enveloping us. We bring our recognition into our reality. Either we create something new, or we continue something we're accustomed to. Everything that we create with our thoughts and feelings has a vibratory quality that returns to us. With our visionary and emotional recognition within ourselves, we can realize our fullest potential.

Living Ineffably While Being Incarnated

Life in the empirical world is inherently limited, while our consciousness is unlimited. As long as we restrict ourselves to awareness within the world of our senses and our mental logic and reasoning, we are unaware of our greater consciousness. We remain ignorant of our inner knowing and higher guidance.

The quantum field, in which we experience our empirical world, is challenging us to realize our greater potential. We can recognize our limitations and decide if we want to maintain them. Keeping ourselves locked into a world dominated by negative energy is becoming increasingly difficult, due to chaos and destruction in all areas of life. We are at a turning point in human history, in which we must choose the way forward, either to remain in the realm of empirical duality, apparently separated in conscious essence from other conscious beings, or we can choose to open our awareness beyond time and space to our participation in unified consciousness. This can happen in our own realization through our intentional desire to expand beyond our accustomed limitations.

By intentionally being aware of our limiting beliefs about ourselves and accepting them as necessary for our true human experience, we can resolve them in favor of awareness of our infinite essence beyond duality. This realization cannot come from outside of ourselves, and there is nothing other than our own limiting beliefs that keeps us fixated on a realm based on fear and belief in mortality. Unless we believe and create the experiences of lack, suffering and death, we cannot recognize, realize and experience them.

Changing our focus of attention to divine love and joy is our great challenge. The world of humanity appears to be self-destructing, and we cannot adequately deal with it. We are being challenged to open our awareness beyond this realm to a reality that is life-enhancing in every way. We can allow the enhancement our DNA into a crystalline arrangement of the molecules of all strands of our DNA. This happens naturally as we open our awareness to our infinite presence.

We are constantly creating patterns of energy with our thoughts and emotions. As we expand our awareness, and we can recognize the qualities of energy around and within us, we can freely choose the qualities that we wish to experience, by paying attention to them and aligning our vibrations with them, and by feeling and realizing their presence. We are free to open our awareness to the universal consciousness that is the essence of everything. Mastery of life is available for all of us.

Working with the Energies of Relationships

All relationships are within our own consciousness, because we are all participating in infinite consciousness. Without our self-imposed limitations, we can be aware of everyone's awareness. Our personal essence consists of our attention. We are each a presence of awareness with infinite creative power. Constantly we create the qualities of our relationships by how we feel about ourselves. As we sensitize ourselves to our own feelings and learn to direct them creatively, we cease being victims of our own ignorance. We can learn to regard every thought and feeling as an expression of the divine, because that is exactly what it is. Every thought and feeling in every moment is creative. It shapes and forms our personal energetic expressions that interact with all of the energetic patterns in our environment, creating our experiences.

When we know how energy flows in our own consciousness, we want to relate with others in the most wonderful ways, because it's the most fun, and we know that we are creating a continuation of the expressions of our heart. When we maintain this perspective, we enhance our energetic presence. All the cells of our bodies fill with vitality and regenerative energies. We can fill ourselves with gratitude, knowing that we are attracting experiences that resonate with our vibratory patterns.

Everyone we meet is an expression of an energetic pattern in our consciousness. All of them are inviting us to be in gratitude, compassion and joy, because these are the feelings that we love the most. If they are not present in situations that we encounter, we can transform the energies in our awareness with intuitive alignment with our heart-consciousness. We can learn to use our imagination in powerfully-focused, transforming ways that change the limiting energetic patterns held in our subconscious.

By living with a life-enhancing perspective, we vibrate ourselves into a world of love, joy and fulfillment in every way. We depart from the world of fear and belief in mortality and enter the world of higher vibrations. We can learn to align ourselves with the consciousness of the Creator, which is our own deepest essence and conscious life-force. We can open ourselves to the unconditional love of infinite creativity and so much more. Everything is available to us, including the most wonderful relationships, because we create them in our own imaginings and feelings in every situation.

Transforming Aging to Ageless

Most of us probably don't plan on living in this body for much more than a hundred years. In reality we are unlimited. From the perspective of ego-consciousness, bound by limiting beliefs, this is inconceivable. Our limitation has been our belief that our cells age and eventually die. No one enforces this belief beyond ourselves. We are free to believe it or not. It reaches deeply into our consciousness and is supported by nearly all of humanity. This belief is so deep and strong, that we need a special way to resolve it. For transcendence we need a strong imagination and the ability to extend our emotions beyond human self-imposed limitations. It requires a similar effort to awakening from a deep sleep to another energetic dimension.

The method is to find the place of awareness in ourselves that absolutely knows anything about ourselves. We can pretend that we have this awareness by imagining what it would be like to live with its qualities. This brings us into alignment with its energetics and into our recognition of it. When we are receptive enough, we can realize that we are in a more wonderful state of Being. We can participate in a higher and more joyful world, separated dimensionally from the limitations of humanity.

It can be helpful to breathe deeply and rhythmically, listen to inspiring music, spend time barefoot in beautiful places in nature, and imagine living in conditions, relationships and energies that we love. In this way we align ourselves with the expansive vibrations of greater awareness. We can become sensitive to the vibrations of our planet and every place and situation in our presence, and we can know the most life-enhancing ways to interact.

This level of consciousness expresses itself in creatively life-enhancing ways that we can feel deeply within and that we know intuitively are perfect. They are our guidance to higher consciousness. When we pay attention to them, their quality of awareness is always present for us, and when we follow their guidance in awareness of our true expansive essence, we can resolve and transcend our limitations. We can become aware of our timeless presence of awareness with infinite creative power. This presence expresses itself in the energetics of life-creation and enhancement. It is what we know is true.

Living in the qualities of the creation of all life and its enhancement, we allow the conscious life-force of the Cosmic Creator to flow into and through our consciousness. There is no death or aging. Our beliefs in Human limitations dissolve from our awareness, along with all fear. Our awareness extends as far and as deep as we are willing to go. This is true for us now and in every dimension. We are free to express ourselves in any kind of physical body, free of any karmic constraints. They exist only in the realm of duality and do not translate into a higher dimension.

Realizing the Truth of Who We Are

Although we are expressing ourselves as humans, our essence is much more expansive and powerful. Awareness of our real Self can come about by focusing on the angelic presence of everyone in our awareness. We are alive only by the consciousness of our heart, all of us, even the dark ones. When we choose to be aware of that presence in everyone, we can realize our deepest connection with infinite consciousness. We can transcend our limiting beliefs about ourselves and realize our essence beyond words and concepts and beyond time and space. In this dimension there is no fear, only mastery of everything.

By paying attention to the presence of our greatest love, first in our imagination and then in our recognition, we can align ourselves with its vibratory resonance, and we can melt our awareness into its presence. Without any input from ego-consciousness, we can become pure awareness, and the consciousness of our heart expresses the essence of the Source of our conscious life-force in its unlimited delivery. We can realize our non-localized presence of awareness without limits.

As we focus our attention on what we want to create, and we hold ourselves in gratitude, love and joy, we can transform our experiences into miraculous expressions. We can become aware of our participation in the consciousness that creates everything, and of our awareness of this consciousness in everyone and everything. Although we have disguised ourselves as humans, each of us is an individual expression of the One Consciousness in all of its fullness. This is what we are being prompted to awaken to.

There are many forms and energies that we can feel good about. It does not matter what they are, as long as we feel aligned with our heart-consciousness, when we focus on them. The original division of consciousness is male and female in balance with each other. Most of our languages recognize gender in everything and everyone in subtle and overt form. The energetic conjunction of male and female in resonance is an ecstatic experience that we can learn to balance in all areas of life.

When we are aware of our essence, we are complete in ourselves and can experience whatever we desire. We can be in fulfilling relationships with our soul mates and having wonderful experiences with them. We can feel each other's presence strongly, even at great distances, because our shared essence is beyond time and space.

Navigating the Split in Reality

Through the international monetary system, we have had our life force drained from us and then used to control, poison, incarcerate, enslave and subject us to every kind of negative energy, and ultimately to stare into the face of pure evil. This is the spectrum of energy that we have engaged with in the dimension of human life. Resisting it is futile, if we want to remain embodied. It is the nature of this aspect of reality. It becomes real for us through our creation of fear, and it allows us to master it through realization of unconditional love.

Fear allows us to engage with negative energy, which would be impossible for us, if we followed the guidance of our heart-consciousness. We can transform a negative perspective and vibratory frequency into a realm beyond polarity in alignment with the creative expressions of life-enhancement. By opening our awareness to gratitude and joy, we can create resonance with our heart-consciousness.

When we can hold this awareness while going about our lives, we become transformative energy centers for ourselves and everyone around us. We align ourselves with the consciousness that creates everything, and that offers us inner guidance continuously. We are aware of this guidance when we pay attention to it with gratitude and joy in a perspective of life-enhancement.

We do not experience negativity unless we intentionally choose to. It is a matter of vibratory resonance. We adjust our energetic signature moment-to-moment with our attention, intention and emotions. Between the vibratory spectrum of fear and the life-enhancing energy of love, we experience a change in consciousness. When we choose love, we no longer need to resist negativity. Having a perspective that is only positive transforms our energetic presence into a higher octave of experience.

As we choose to align with the life-enhancing energy of our heart, we change the energy of humanity. We cannot participate fully in the human experience without knowing fear, but we no longer have to subject ourselves to it. We are playing a game of consciousness, in which we direct the qualities of our experiences by our own vibrations. We can learn to be intentional in every moment, creating everything we need, while also creating the energy of transformation for the Earth and all of humanity. By the vibratory qualities of our thoughts and emotions, we express our state of being, forming our energetic signature, which determines our presence in the energetic spectrum of love or fear. Everything in our experience changes according to the one we choose.

Living in the New World

As we awaken from the hypnotic trance of 3D humanity, we may need to reorient ourselves. It's like awakening from a very deep lucid dream. We have known a personification of ourselves, but we have not known our true essence. We have lived under hypnotic limitations and cannot easily transcend them. Each of us has our own way of accomplishing this. A strong intention and asking for help from our guides and angels can open our awareness. Deep meditation can take us beyond our presence in space and time. Becoming aware of our inner sound current and immersing our awareness in it is another path. What we need is a kind of reverse hypnosis in which our limiting beliefs dissolve, giving us absolute freedom, fulfillment and unlimited awareness in the consciousness of our heart.

We can become intentionally aware of the light in everyone. It is a reflection of our own essence, being presented to us for our acceptance, compassion and love. As we become able to fill ourselves with these vibrations in any moment, we expand our conscious awareness beyond the limitations of the judgmentalness of ego-consciousness and into the unconditional love of creator consciousness. This elevates and magnifies our vibratory radiance. When we feel our eternal essence, we can identify with an energetic spectrum that we love completely and eagerly, because it is innately true for us.

When we achieve this awareness without attachment or ego-involvement, we open the access to our hidden abilities of infinite creative power and awareness. We can realize that we are not mere humans, subject to powers outside of ourselves. All of us, regardless of whatever condition we are in, are where we are of our own volition, which we can learn to direct. As expressions of Creator Consciousness, we have absolute power over our lives, circumstances and experiences, once we realize the truth of our essence.

We can awaken to a world of beauty and joy by recognizing its reality. It already exists right where we are, waiting for our realization in alignment with its vibratory resonance, which we can know in the consciousness of our heart. It is a consciousness that we all share in the enhancement of all life. Beyond suffering and death, it fills every cell in our bodies with vitality. Wherever we direct our attention, we can experience grateful and joyous inspiration and realize their reality in our awareness.

Achieving Transcendence

Especially in the cities, much of life has become life-diminishing and depressing. Anyone who continues to be entranced in 3D life faces a great challenge to break out of it into a higher energetic dimension that is life-enhancing in every way. It is becoming more difficult to make the leap in consciousness from the world of duality to life filled with love and joy. The world of duality has been taken over by the dark force and is becoming increasingly intolerable, but there is another choice that we have for our experience.

Because everything in our experience is an expression of consciousness, we can use our consciousness to create what we want. For this we need the cooperation of our subconscious, as well as the consciousness of our heart. We can control the quality of our awareness with our thoughts and emotions. What we want to pay attention to and how we feel about it is a matter of personal choice. These are the vibrations that we are emitting into the quantum field, and they will be brought into our experience.

We live in a field of energy filled with potentialities that are subject to our energetic radiance. None of this is obvious to ego-consciousness. which means we are unaware of these energetic mechanics. As a result, most of us live as if life is haphazard, and we get what we want by physically doing things. We have not realized that 3D life is a game of consciousness that we are all playing together with rules of limitation that we have imposed on ourselves, and that we can become aware of and release for our personal transcendence.

As we realize that we are in a play of consciousness, we can learn to change our circumstances and quality of experience. Because every potential experience is available to us, we attract into our lives the qualities that we pay attention to and feel. If we do not realize that we are creating our experiences in every moment by the way we are, and by our perspective and reactions to what we're aware of, we give up our intentional creative direction. We create our experiences subconsciously, according to our limiting beliefs about ourselves, locking us into the 3D trance.

The path to awareness of our true identity and abilities is to change our vibratory presence to align with the life-enhancing energy of our heart-consciousness. It is our connection to the unconditional creative love of our greater Self. This opening to expanded consciousness provides a deep knowing and feeling of life-enhancement in every way. It enables us to live in a dimension of gratitude, love, compassion and joy apart from a world of negativity. It happens without any outward action on our part, and it fills us with everything our heart desires.

Creating the Life We Desire

In the past, society-wide tragedies elicited fear and anger among us, but times have evolved. Now tragedies elicit increasing amounts of compassion for those who suffer as a result of what happens. Fear grows weaker. We have grown accustomed to threats and intimidation, and the power of compassion is weakening the hold of negativity upon our consciousness. Even without our being aware that it is happening, the vibrations of humanity are rising and becoming more positive with a greater intention to care for the Earth and to brighten our lives.

There is a realm beyond polarity that exists right along with our world of duality. The point of separation is the thin line between fear and love, between diminishment and enhancement of life. In every moment, we must choose either one as our state of being. We are essentially loving or fearful. Most of us cycle back and forth. As we become more positive, we become heart-centered and moved by gratitude, joy, compassion and love. These are the emotions of our essence, our soul and greater Self. When we feel their presence within, we are transcending our ego-conscious self-limitations. Through our emotions we can feel the energetic structure of our experiences. We can feel the radiance that we emit into the quantum field, attracting resonating energy patterns that become our experiences.

Once we become aware of the cause of our experiences, we can take control of our personal lives and direct our experiences as we prefer. Anger and fear are life-diminishing. Joy and love are life-enhancing. We choose which one we resonate with. That is where our attention goes in creating the qualities of our experiences and determines which dimension we live in. Our state of being does not have to interact with the energetic patterns of experiences in our outer environment. We can choose our own vibratory presence. This is what is important in creating our experiences, not reactions to outside stimulus.

Although we have believed the empirical world is physically solid and composed of particles of matter, quantum physicists have shown the smallest constituents of our world are actually rapidly spinning conscious energy centers that appear to us as particles. Physicists track them as wave patterns. Our bodies and all material things are in reality complex interacting wave patterns of electromagnetic energy, having amplitude, frequency, wavelength and polarity. All are expressions of consciousness. The vitality of every cell in our bodies is an expression of our consciousness. Our state of being is how we feel about ourselves. This expresses our energetic signature as the quality of our radiance.

Everything is conscious and part of infinite consciousness that has its own identity. We can call it the consciousness of the Creator. It is the One who determines our innate essence as an aspect of Itself, giving us access to infinite awareness, which we have self-limited in order to play in the human experience of duality. As we become aware of our unlimited Self, we enable our mastery of all worlds through our ability to control the vibratory alignment of our awareness.

Directing Our Mental and Emotional Alignment

When we sit and enter a state of receptivity, many things become clear. When the ego is relaxed and non-intrusive, we can open our awareness to energy patterns beyond the limited consciousness of the ego. We can enter the realm beyond thought, into the state of knowing. This is a knowing that the ego cannot disprove. The presence of knowing includes the emotions of the heart. We are in a quantum state of limitless, life-enhancing creativity. We can experience non-localized gratitude, love, compassion, freedom and joy. This is the beginning of Self-Realization.

This comes after much practice of maintaining a presence of compassion and gratitude in the face of every experience, even the darkest encounters. We can use our imagination to elicit the ability to see the light of Creator Consciousness in all beings. This is the vibratory presence that we want to be in alignment with, and it will deliver the quality of experience that resonates with our present vibratory signature.

As we become aware of our senses beyond the physical world, our awareness begins to open to a brighter, more beautiful world of experience, in which everything is true. There is no dissimulation. It is toward this kind of realization that we are being called. It is innate within our own being and is what we deeply know. In this state of being, we can be aware of ourselves as living fractals of infinite consciousness. The expression of our human persona is directed by our complete consciousness, according to our desires for certain experiences.

Awareness of our presence beyond our human persona enables us to understand what human life means for us, and why we experience the scenarios that we do. This is part of our inner knowing. When we are in a state of clarity, our hidden abilities appear in our realization. If we can realize that we are infinite consciousness, there is nothing that can keep us from manifesting anything that we desire, and we can be thoroughly heart-centered in all of our motivations. We have absolute freedom to enjoy our best lives in deepest love and gratitude.

On the way to Self-Realization, we become aware of our infinite essence and unlimited creativity. We can realize our eternal essence of conscious presence with unlimited abilities in creating our personal experiences in whatever way we wish to express ourselves. We only have to focus on what we want and align our emotional presence with it. Whatever we experience in any dimension is a result of our own energetic alignment.

Going Deeper into Self-Realization

On the inner path to Self-mastery, the most important process we must go through is getting ourselves into a high-vibrational mental and emotional presence, beyond limiting beliefs about ourselves. This happens when we intentionally open our awareness through our imagination, creating visions of beauty and feelings of gratitude, joy, compassion and love. We all know what this feels like, because it is our natural state of being. By using our imagination, we can open our receptivity to the greater consciousness of our essential Being. We know when we're in a state of gratitude and joy.

When something about ourselves isn't quite right, we can feel it. Whatever keeps us from being in joy can be identified and examined. If it arises from any kind of fear, it must be resolved through resonance with the consciousness of our heart. As we intentionally adopt a perspective of life-enhancement in all ways, we open our awareness to the light in every heart and the presence of infinite awareness. We gain awareness and use of our higher abilities and creative power.

Since we are individual extensions of the consciousness of the Creator, in our essence we are unlimited. We are each a center of conscious presence of awareness with unlimited creative power. We can express ourselves as anything or anyone we desire to imagine. The only thing keeping us from this realization is our system of limiting beliefs about ourselves. They have been imposed on us telepathically and enforced by society since birth and beyond to family inheritance. Because they are so deeply embedded in our consciousness, they require great determination to penetrate and resolve.

We can begin with a desire to know our true nature. We can pay attention to our inner prompting in every situation. This may require intentional practice, since our limiting ego-beliefs do not allow us to recognize our inner knowing. We must search for it within, by imagining how we can express the consciousness of our heart in every situation. By directing our imagination to create life-enhancing scenarios in every situation, we align ourselves with the vibrations in the quantum field that manifest those energetic patterns in our experience.

Ego-consciousness is limited to thinking and sensing the empirical world. It cannot know what the heart knows, because heart-knowing is beyond thought. Through mental and emotional direction, we can become aware of our inner knowing. It is a matter of adjusting our vibratory presence through our mental and emotional processes. We are actors within our own consciousness, and we can identify with any energetic patterns we choose. When we accept and realize the reality of our infinite awareness and creative power, we can fulfill the destiny of our divine essence.

Using Our Emotions for Transformation

Every vibration has an identifiable feeling. By controlling our emotional state, we are choosing the vibratory spectrum that we wish to experience. We choose every moment, either consciously or sub-consciously. Through intentionally entering high-vibratory imaginings and feelings, we can train our sub-conscious to come into alignment with us. By holding ourselves at the vibratory level that we love and are grateful for, we begin to feel those vibrations around and within us and in others around us. As our outer experiences come into resonance with us, the others around us may change, as they are drawn into our vibratory resonance.

If we are experiencing the limitations of ego-consciousness, none of this makes sense. If we want to understand and realize a truly wonderful life, we must resolve our fear of suffering and death. These are the basic beliefs that enable our ego-consciousness and support all of our limiting beliefs about ourselves. Giving them reality for us happens in our attention and vibratory alignment. By changing our awareness to energetic patterns of gratitude and life-enhancement, we stop giving our life force to negative energetics.

By creating and staying in a vibratory range of gratitude and joy, we can become aware of our inner prompting and knowing. This is the presence in our awareness of deepest feeling, knowing and understanding. We can be aware of our essence beyond time and space, as a non-localized and individualized presence of awareness, arising within infinite consciousness. This awareness has unlimited creative power that can be used in alignment with the vibrations of our heart-consciousness. When we're in this alignment, we feel this connection deeply.

As our inner awareness clarifies, our intuition becomes our primary and constant guidance, connecting us to the consciousness of our guides and angels and the consciousness of the Creator. Living as an expression of our greater Self results in mastery of the 3D world. By living in alignment with our deepest inner knowing, we can always consciously choose to feel ourselves in a state of gratitude, love and joy, giving us a powerfully radiant presence and a life that is fulfilling in every way.

Currently we are spanning two worlds—the positive and the negative. Our free will allows us to choose either polarity, and we do, in every moment. We can be observers of the negative, but it's been customary for us to get sucked into a fixation with it. Staying in gratitude and joy is the transformative energy that carries us into a higher dimension in our present awareness.

Transforming Time and Space

In the empirical world of time and space, space is 3-dimensional and time is linear in 2 dimensions. Because everything that exists is an expression of consciousness, we create our realization of time and space in our own consciousness. In our experience, it does not exist outside of our consciousness. We can change our realization of time and space, which will change it in our experience. The next step in the ultimate realization of infinite awareness, is a different idea of time and space.

There is a realm of experience that has been created in the conscious realization of humanity where every experience is constantly instantaneous for us, depending on our energetic alignment. Everything is possible. Space is less dense than in the empirical world and more striking in illumination and in feelings of life-enhancement. Everything feels really good physically and emotionally. The air is more fluid. The water is lighter and brighter. Love and joy fill each moment. The air and the water caress our bodies in pleasure. Life is fulfilling in every way. Creation and experiencing are instantaneous in constant alignment with our mental and emotional processes.

Moving beyond the limitations of our mental processes, we can enter a realm of knowing. Instead of feeling separated from everyone we can enjoy telepathic union and empathic sharing. Being constantly aware of our intuitive knowing, and following what we know, we can be masters of our lives. In the deepest knowing of our heart-consciousness, every aspect of our bodies speaks to us, giving us guidance that carries us into a life filled with joy and beauty. Everything in our experience is calling us to greater awareness of our true essence.

As we learn to pay attention to our inner knowing in the consciousness of our heart, we gain confidence that it is always true. All of our encounters become symbolic forms of communication from within, enhancing our awareness. The qualities of all of this depend upon the energetic resonance of our attention.

At any time we can enter an expanded energetic spectrum in our awareness by our intention. A higher-vibratory realm of experience already exists in the quantum field, which envelops us. It only requires our recognition and realization through our imaginary and emotional vibratory alignment with the consciousness of the heart of our Being. In deepest gratitude, love and joy, we create its reality in our experience, as we align ourselves with the Consciousness of the Creator.

Realizing the Importance of Our Attention

In our natural state of joy, we are intuitively given understanding of every vibratory pattern that we pay attention to. Every situation becomes clear as we feel the qualities of its energetic patterns. When something doesn't feel good, it's time for our personal transformation. What we perceive is a reflection of our own consciousness. What we feel is our own emotional state of being. This is controlled by our focus of attention and the vibratory quality that we choose to feel and pay attention to. By imagining and feeling scenarios of gratitude, compassion, love, joy and their companions, we create an energetic radiance that draws everything that is life-enhancing into our experience.

As we begin to live in the realm of love and joy, our physical experiences come into alignment with us. Instead of reacting to circumstances in our experience, we have the choice of creating a transformative state of Being that we love and are grateful for. The more we do this, the more we train our subconscious mind to help resolve our limiting beliefs about ourselves and to allow ourselves to live in complete freedom and abundance. This is our natural state, because we are fractals of infinite consciousness and creative ability. We can be however we want by aligning with the desired energetic qualities in our imagination and feelings.

If we imagine ourselves living in wonderful and fulfilling

experiences, we create a resonant frequency that interfaces with such experiences in the physical world. Our lives fill with loving and cherishing relationships. As we learn to release our doubts about our creative abilities, our awareness expands into greater consciousness. Our understanding of life deepens, and we can train ourselves to give our attention to energies that are supportive of all life in every way.

By aligning our mental and emotional processes with our heart-consciousness, we create an environment of energetic renewal for the Earth and all of humanity. This is further strengthened by the expanding life force of the Spirit of the Earth. This energy has no interaction with negativity, leaving those who are negatively-polarized to become unstable and insane, as their source of life force dissolves. This creates chaos in society, stimulating greater realization within human consciousness.

As more of us realize our emotional magnetism and the importance of the energetic quality we pay attention to, we can bring our personal, as well as our larger environment into alignment with life-enhancing energies. In order to realize our powerful creative ability, we must resolve and transcend our limiting beliefs about ourselves and truly know ourselves within the consciousness of the heart of our Being, becoming a transformative force in the consciousness of humanity.

Entering a State of Transcendence

As the realm of duality dissolves out of our awareness, we can empower a world of love and support for all life by focusing on it and feeling ourselves in alignment with it. It is an expression of the heart of our Being and is our natural desire, once we have released fear and limitation from our lives. We can live in a world of joy and fulfillment in all ways, where our currently-hidden abilities come into realization.

Through programming to worry and be afraid of suffering and death, we have learned to be the gatekeepers of our own consciousness. Subjection to negativity is not a necessary condition. If we choose to recognize our limitations and examine them down to their roots, we find that they are all self-created and have no validity beyond the realization that we give them. Only if we give them reality through our mental and emotional alignment can they come into our experience.

On the other hand, if we choose transcendence into the realm of love and joy, we can realize fulfillment of every desire. As fractals of universal consciousness, we are the creators. Everything we experience is brought to us by our energetic resonance within the quantum field that envelops us. We can feel energetic vibrations, and we are free to identify with any of them. Because we have been thoroughly trained to stay in constant worry, fear and doubt, they prevent us from being aware of our eternal essence and infinite awareness beyond space and time.

Our experience in space and time is artificially real. Its reality is created in our own consciousness in interaction with the empirical range of electromagnetic energy patterns in the quantum field of all potentialities. The entire empirical world is nothing more than swirling patterns of energy that our consciousness interprets as real for us. Because of our creative essence, we make it real for ourselves.

We do the same thing with positive and negative energies. They are only potential until we align with their vibratory frequencies and polarity. When we recognize them, we can realize them as real in our experience. By changing our realization, we can change our reality. For this we can use our imagination and emotions to create a state of being that opens our awareness to our true inner knowing. Once we achieve intuitive alignment with the consciousness of our heart, we enter a realm beyond fear. It is a realm of deep understanding and transcendence beyond limitation.

Disempowering Evil Control of Humanity

As we face the horrific tragedies and challenges in the war against humanity, it behooves us to know that nothing happens by accident, and we have the choice of how we interact and feel about what is happening. Although we share the energy spectrum of positive and negative duality with humanity, it is important to realize that everything we experience consists of energetic polarities that we either align with through acceptance or resistance, or we transcend and transform.

Evil cannot be defeated by identifying, judging and fighting against it and seeking revenge. When we engage with it in this way, we must align with its vibratory level. It is the nature of this energy to destroy life. By fighting evil and those who identify with it, we actually empower it with our life force through our anger and fear. Fighting evil with the intent to destroy it only strengthens the energetic patterns that create it, and we pay a price in diminishing our own life force.

There is a much bigger perspective that we can realize. Nothing happens by accident to any of us. Every bit of suffering and death that we experience is part of our life plan and unfolds in alignment with our own vibratory state of being. All of these things are lessons for us in learning to realize our own higher guidance beyond the realm of duality. In our ego-consciousness, we are victims of our circumstances, but in our true selves, we are masters of our situations. Our transformation occurs in our own realization by resolving our limiting beliefs about ourselves and transcending them.

When we are living in victim-consciousness, we need saviors to relieve us of our afflictions, because we do not know that we are the cause of our own condition. The intensity of our own experience depends upon our strength of devotion to the diminishing or the life-enhancing qualities of our intentions. When we are clear mentally and emotionally and aligned with our deepest inner knowing, we can transcend limiting beliefs about ourselves. Our environment reshapes itself in alignment with our predominant energetic polarity.

Because everything arises in our own consciousness, we have the ability to choose the vibratory qualities that we pay attention to. These are the energy patterns that we resonate with and that magnetically attract new experiences for ourselves. They resonate with the qualities of our attention. By living with higher guidance, we elevate the consciousness of humanity toward Self-Realization of our true Being.

We can face negative experiences with the guidance of our heart, which is fearless. As humans, we do not know our destiny, but we can anticipate knowing how to handle any encounter. When we realize that we are our eternal presence of awareness, we can understand our life completely. Aligning with ego-consciousness in the desire to engage with negative energy to destroy it, results in the enhancement of life-diminishing energies. By constantly living with the guidance of the heart of our Being, our hidden creative abilities activate, and we can no longer be intimidated.

Realizing Infinite Awareness

When we are aware of the qualities of energies within and around us, we can be aware of the lessons that confront us. We express ourselves through our physical body in an empirical world that is constantly made real by the beliefs of humanity. We emit electromagnetic energy patterns through our mental and emotional processes, and we direct our lives by what goes on in our thoughts and feelings. These are the energy patterns that we are constantly creating and imbuing our experiences with. By being aware of the vibratory quality of our thoughts and feelings, we can realize that we can control and direct them.

Everything is energy, and we express this energy with our consciousness. The physical world is what we imagine it to be. Within our consciousness, everything exists, and we can be aware of everything that we allow ourselves to know about. It comes to us through our intuitive knowing in the depth of our essence. It is the conscious life-force in the heart of our Being. When we recognize it, we realize that we know it well. It is our own presence of awareness beyond space and time.

Realizing that we are living beyond space and time concurrently with our role as humans, we can understand the entire complexity of the human experience in encountering negative energy. In order to take this experience as real, we had to create our limiting beliefs. When we desire to realize our true Self, we must know that we are not required to maintain limiting beliefs. We are fractals of the universal consciousness of the Creator, making us infinite in our presence of awareness, and with free minds and emotions, enabling us to create unique patterns of energy in the constant expansion of consciousness.

Because we've realized every quality of evil over many lifetimes, we no longer need our limiting beliefs about ourselves. We can resolve and release them and open our realization to our infinite essence. It is a most freeing experience to understand the magnitude of who we are. Working with our thoughts and emotions in intentionally creative ways is what we're designed to do. We can't not create. It's our nature to create the energetic feeling of every experience in our lives.

We get to choose the quality of every experience that we have. When we don't realize what we're doing, and we don't intentionally choose our thoughts and feelings, we believe that life is haphazard, and things just happen. Once we open ourselves beyond our limitations, we can master ourselves and create experiences that enhance life everywhere, including our own.

Compassion and Love Are Rising

The consciousness of humanity is becoming more positive, more compassionate and loving, while also realizing the great evil that we are subjected to. Wide-spread distrust of the government is growing, and people are banding together to improve life apart from corporate tyranny. The fires in Lahaina are a good example of a tremendous amount of compassion and love that is arising. The nearly-complete destruction symbolizes the dissolution of an old way of life, and invites the arising of a new culture and civilization based on love.

The act of war against humanity resulted is great tragedy for those who lost their homes, businesses, loved ones and pets, and their way of life. but the tragedy is ending, and the survivors, who have nothing, not even a car and probably not a job, need us to act in unity with them, and we are. They are not victims; they are survivors. More than 11,000 of them are being housed and fed in the resort hotels. Transportation is being provided, and money is being distributed. Every service they could want is being made available. These are the physical manifestations of compassion and love.

More importantly, the energy of compassion and support has risen world-wide. People everywhere have been praying, and some are imagining something wonderful arising from this event. The something wonderful is a new culture of creativity and love, manifesting in a beautiful masterpiece of a town. Maui is too wealthy and enlightened for this opportunity of transformation and renewal to be missed. The only significant things remaining in Lahaina are some coral-stone brick walls and the giant banyan tree. The town has been reduced back to nature with a potentially beautiful and amazing future.

It's not just the physical things that are happening, but they are the manifestation of the energy that we are creating and radiating. This is the beginning of the era of gratitude, love, compassion and joy. The survivors know that they are being cared-for out of love and compassion. This elicits a conscious background of gratitude and a rising amount of joy as the grief lessens. Most of the support is given without expectation of anything other than knowing that they are supporting those in need. And it's not just temporary. This is a long-term process. Everyone involved knows this.

On a more expanded level of consciousness, we can know that all of these things are having an effect on the consciousness of humanity, as well as on the individuals involved. The feeling of all being in the same family is rising into awareness. As we continue to align with compassion and love, we become aware of our true inner knowing, because we are not distracted by any personal needs or fears. All of those get magically covered. To the ego, this is impossible, but it happens, because its' a logical energetic result of how we modulate energetic patterns with our mental and emotional processes. In this case those are symbolized by compassion and love in the essence of Divine Consciousness.

Realizing the Essence of Our Consciousness

When we finally open our realization to our real identity, we can dramatically change our lives. Having our awareness confined to the hypnotic trance of humanity upon incarnation, we are involved in a process of Self-discovery. When witnessed from within, the dualistic empirical trance is very convincing and attention-consuming. If we are to expand our awareness beyond this trance, we must look within our own consciousness for greater realization.

Since infancy, we have been programmed by our society to adopt beliefs about ourselves that are severely limiting, making us create our own victimization, poverty and loneliness. None of this happens outside of our own consciousness. As a result of our beliefs, which control our mental and emotional processes, we are completely self-responsible for everything we experience. We have been our own enslavers.

As we proceed along the path of expanding our awareness, we can increasingly be present without ego-involvement. With all of its unfulfilled needs, the ego is the expression of our limiting beliefs. If we are to release its hold on our awareness, we must recognize its limiting presence. Without higher guidance, it has assisted us in navigating our life experiences. But we are capable of so much more.

We have many clues to guide us, and we have amazing abilities, waiting for us to realize. We have a guiding consciousness that we can feel and resonate with. It is the conscious life force that is part of our essence. When we are in alignment with it, it expresses itself as creation and life-enhancement. It fills us with joy, freedom and self-fulfillment. We may identify it as the expression of the heart of our Being. It is symbolized by our physical heart, which lives to give us life within the limits that we create for it.

As more of us awaken to these realizations, we expand the consciousness of humanity, making it easier for everyone to realize the greatness of who we are. We can all realize the emotions of our soul, such as gratitude, joy, love, abundance, freedom and limitless life. These are part of our true Being. When we search for them, and use our imagination, we allow them to come into our experience. It is only when we are open to them and desire them without doubt that they can manifest for us. Because we are so deeply imbued with doubt and fear, we must be able to surmount our programming through a strong desire and intention to expand our awareness. This expansion happens naturally when we come into resonant alignment with the energetic expression of our heart.

The Value of Pain and Suffering

Pain and suffering tell us what we don't want to know about ourselves. They tell us where we're enamored of negativity and how we're mentally and emotionally stuck in life-diminishing energies. Pain in our bodies symbolizes the kind of negative energy we're attached to. Our ego-consciousness does not want us to realize the ways we limit ourselves, but there is an aspect of our consciousness that wants us to have this awareness, and it prompts us with pain and suffering to open our awareness. In response, the ego hangs on tighter to our limitation and blames others for our condition, or we believe that everything is a matter of chance, fate or karma.

Often it may appear to us that others are the cause of our suffering. This would be possible only if we are victims, without control of everything we experience. If we actually are sovereign Beings and expressions of the consciousness that creates everything, then we are the source of our experiences, and it is possible for us to realize how we have been creating our conditions. Our suffering always has a specific form that symbolizes where and how we are holding a negative fixation. In her book, *Heal Your Body*, Louise L. Hay has presented a comprehensive review and analysis of all the ways that we disempower ourselves with our physical aches and pains, and how we can take positive control of our lives again.

Opening our awareness to our physical problems and recognizing their symbolic significance for our emotional distress can be the path to healing ourselves. It requires opening our heart in compassion and forgiveness for ourselves and everyone we blame for our condition. We have manifested our aches and pains from our fears and judgmentalness. If we can resolve and release all of this, and see the light of the Creator in everyone involved, we can dissolve our trauma, accept our situation as our own creation, and transform our perspective with confidence and love. We can move forward with joy in our lives, as we feel the guidance that comes to us through the vibrations of our heart.

Living with pain and suffering is not necessary. It is all self-created by dwelling on some form of negative energy that we hold in our bodies. Anything that gives us discomfort deserves our attention. We would be experiencing it only if we have invited it into our presence with our vibratory resonance. It does not matter how this came about, whether we inherited it or created it in this life. If we continue to hold onto it and do not realize the effect it is having on us, our deepest consciousness provides physical defects that prompt us to pay attention to the thoughts and feelings that keep us from radiating love and joy in the presence of everyone and everything.

Living Intentionally in the Field of Consciousness

We have believed that we are separate individuals with our own private thoughts and feelings that we can keep to ourselves, even though we have known that there are telepathic individuals who can read our minds and emotions. Now we're finding out that there is technology that can do the same thing, and in some cases, even influence and direct us to do things, even things that we would never want to do on our own. This implies that our consciousness is not localized and that we live in a field of consciousness, within which we are each a focus of attention with a unique identity, which we may recognize as an energy signature, or an electromagnetic expression.

Although we are enveloped in a field of every possible thought and emotion, there is a mechanism that gives us our own thoughts and feelings. It is the polarity and vibratory frequency of our mental and emotional processes, both conscious and subconscious. These are subject to our personally-chosen focus of attention. We are free to subject ourselves to random thoughts and feelings in our accustomed state of being, and to react to outer circumstances as if they are important, and they establish our energetic identity, which we broadcast into the field of consciousness through our personal radiance.

Our energy signature is not a secret. It is recognizable to all who choose to perceive it, giving away the qualities of our thoughts and feelings. Our vibratory presence is also responsible for attracting the patterns of energy that come into our experience and repelling those that we are unaware of. We can allow this process of living to happen by default, and we can also choose to direct it as we desire. We have a largely-unrecognized power to create and shape our lives by our own attention and how we feel about it and about ourselves.

We are subject to influences beyond our personal awareness only to the extent that we are open to them. Every personal experience has its primal formation in our own consciousness. if we strongly desire, we can find inspiring ways to clear our mind and emotions and to feel inspired and creative. We can open our awareness and ask for transcendence in alignment with our heart-consciousness and in collaboration with our unseen guides and angels living in the universal field of consciousness. The only blocks within our conscious awareness are self-imposed and can be self-transcended. We are being invited to realize our essence as we experience ourselves as our unique, eternal and unlimited presence of awareness within the consciousness that creates everything.

Our Destiny and Our Choice

As we begin on the inward path toward Self-Realization, there are things that we learn along the way. We realize that human life is a play in consciousness, and all of us are playing our pre-destined roles for ourselves and for everyone else. Every situation and encounter has meaning for us to recognize. If we pay attention, our inner guidance gives us understanding. At the same time, we can shape the qualities of our experiences with our intentional thoughts and emotions, both conscious and subconscious. Because of this, everything we experience is self-created, by our doubts and fears as well as our transcendent presence. We choose the energetic quality we will participate in and experience.

We are all interconnected in consciousness. Each of us is our own unique person with a deep connection to the consciousness that creates everything. In order to play an authentic human role, we must create a mind-set with parameters to our awareness. As long as we stay in this mind-set, we cannot know our true essence, but we can find clues to direct us deeper within our presence. These clues can be anything, and they are directed to us in ways that each of us can recognize.

In our essence, we are all Masters of every aspect of life, but in our ego-consciousness we cannot realize this. If we follow our inner clues and guidance, always searching for the most life-enhancing thoughts and feelings, we can realize a world that we can inhabit now by directing our imagination and feelings to living in the realm we truly want. It is the vibratory polarity and frequency of our perspective and state of being that attracts our experiences, and we can intentionally control this.

We can learn to control our thoughts and emotions in ways that inspire and stimulate us with gratitude and good feelings. We can disengage entirely from negativity, and instead follow the wonderful desires of our heart. As we clarify our mental and emotional processes, we can focus more acutely on what we know is life-enhancing for all of us. Everything exists in our own consciousness, and we can open our awareness to this.

Without limiting beliefs about ourselves, we can realize our true nature beyond time and space and in other forms and persons, as well as our human presence. It is all available for our awareness and realization. When we realize the reality of someone or something, it becomes real in our experience. This realization comes from our intentional opening to it and feeling its energetic presence. We can intentionally create our own reality in interaction with the consciousness of humanity.

Examining and Enhancing Our Inner Life

By resting in the most wonderful energy that we can imagine and feel, we can begin to align with the consciousness of our Creator. Although we may be aware of more, we hold in our consciousness what we perceive within the vibratory limits of this empirical density and energetic dimension. Awareness of these limits is the first step in expanding within our greater consciousness. Next, we can examine our limiting beliefs about ourselves and discover their origin.

In its essence, matter consists of identifiable patterns of energetic vibrations. The dark, anti-matter world we have inhabited is based on fear of termination, while becoming decrepit and moving intentionally toward it. It is filled with entropy and is ultimately self-destructive. It is the realm of ego-consciousness.

Light matter is limitless and life-enhancing. It is what we truly love, identified as the energy of our heart. When we come into its awareness, we are on the way to exploring infinite consciousness and transforming our human presence. In every moment we have the choice of aligning with one or the other by what we pay attention to.

In this life, we are all actors providing energetic encounters, as we learn the best ways to use our attention for greatest fulfillment for the benefit of all. In fearless, love-filled living, the greatest abilities we can use are our attention, imagination and realization. When we learn to direct them in ways that we deeply feel are life-enhancing, we can live in the world of boundless light matter, with fulfillment in every way, even beyond our imagination. In this finer world we can utilize creative abilities we didn't believe we have, because we are able to trust ourselves implicitly to abide in the energy of our heart's intuitive knowing.

Being aware of the light in everyone, we can very much enjoy mastering our lives and our world in gratitude, compassion, love and joy. Intentionally paying attention to these wonderful energies within ourselves, regardless of what would distract us, we shape the world around us and within, including our physical presence. When we maintain a high vibratory presence of joy, we are aligning with the qualities of our conscious life-force, and our awareness opens to a beautiful and heart-felt realm of living.

Imagining and Realizing Our Reality

When our loved ones die, we have the choice of mourning them as gone forever, or we can feel in alignment with their conscious life-stream and follow them into a loving dimension. When we feel complete with this experience, we may return to our limited self, inspired with a new understanding of unconditional love. This is an awareness that we all innately have, once we release our limiting beliefs about ourselves and our mortality.

Releasing is not a strictly mental process. It is in the realm of our deepest emotional feelings. They may not have a form or description, but we know them. Originating in the presence of our conscious life-force, they come to us when we are open to them and desire their presence. This is also where we can realize our own presence of awareness beyond time and space.

We are Who we are, and we cannot change our essence, that is our presence of awareness. We can, however, change any of our expressions, including our human person and how we think and feel. We can change our perspective by opening our awareness in gratitude to what gives us joy, compassion and love. At this point in our realization, we no longer need to engage with negative energy of any kind. Our knowledge is trustworthy and true when we are aware of its essence through its interaction with our own radiance. By being receptive to our intuitive knowing, we can align ourselves with its vibratory resonance.

Awareness of our deepest knowing and feelings without any interference may require searching for them. They enlighten us with everything we need to know in every moment, so that we may express ourselves with the life-enhancing energy of the heart of our Being in every encounter and circumstance.

When we are grateful for our essence and desire to expand our awareness through the vibratory resonance of our heart, we can realize abilities that we kept ourselves limited from. One of these is our creating an energetic pattern or structure in every moment. This becomes the quality of our experience. Whether we are reacting or transforming energetic patterns in our attention is unimportant. It is the quality of our expression that is important. Our energetic state of being is our creative expression.

By being in a state of gratitude, compassion, love and joy, we can live in a less dense reality that is closer in alignment with the consciousness of the Creator. It affects every aspect of our lives and elevates our experiences, apart from the world that has gripped human attention for eons. We are playing a game for control of our consciousness, and we always have choices about what we pay attention to and align with emotionally. This is our creative process and the source of our realization of who we are.

Deepening Our Self-Understanding

As we and our society have programmed our subconscious deeply to keep ourselves entranced in the dualistic empirical world, we have accepted living in a constant background of fear. We have believed that we can be victims of intimidation and other thefts of our life-force, but nothing beyond our own consciousness enforces this entrancement, and we can release our awareness intentionally, by using our imagination and emotions creatively. We have an innate connection with much greater consciousness that opens our awareness to our true essence, our limitless, eternal presence of awareness with infinite creative power. Who we are is so far beyond our human imagination!

Nevertheless, we can have an impression of our essence, and that is sufficient to transform our perspective about everything. We are learning how energy is expressed and how it interacts with other energetic complexes. In our essence we have no darkness or negativity. In our awareness, we arise within the consciousness that creates and enhances all life. It gives us complete freedom to pay attention and align with any kinds of energies we desire to experience, which is how we got into our present situation.

Everything we experience is carrying us into deeper Self-Realization. There is a distinct difference between living in the negative aspect of duality, and living in love and freedom. We innately know what this is, and we can be constantly aware of our own polarity and vibratory quality by how we feel about ourselves. We can train ourselves to feel ourselves living in a world of gratitude and joy, beyond the reach of negativity.

Whenever we are aware that we're paying attention to something negative, we can take a few deep, slow, rhythm breaths and align the direction of our focus with our heart-consciousness. This opens a realm of life-enhancing energies and experiences. By practicing being aware constantly of our inner knowing, we open our awareness to higher guidance. As our awareness opens, we can realize the nature of the play of consciousness that we're living in, and how we can transform our experiences.

Everyone in every encounter in essence is a Being of Light. When we pay attention to the inner light, we can be in alignment with the consciousness of the Creator and can feel ourselves enveloped in unconditional love and joy. Negative energies and experiences dissolve out of our lives from lack of our attention. Our realization can elevate and intensify our radiance, creative ability, and experience.

Impressions of Our Expanding Consciousness

We are designed to want to experience, as fully as possible, the qualities of freedom, joy, love and abundance. These are the emanations of Creator Consciousness, and we are free to choose to experience their fulness in any moment. When we do, we feel good. When we choose to feel some other way, we are given resonating experiences. If there is even a slight tinge of fear, we feel it, and if we hang onto it, we convey our life force to the realm of negativity, making it real for ourselves.

By training ourselves to resolve and transcend our ego-needs, we are able to be present in our awareness and to let all experiences flow through our consciousness, while choosing the ones we love and aligning with them. This keeps our vibrations high and positive, and it enables us to be clear in our understanding. We can open our awareness to our inner knowing. It guides us into fulfilling experiences that we love and enjoy.

Everything we don't like is an energetic pattern that we are holding in our consciousness. All experience is happening in our own consciousness. Everyone we encounter is an aspect of ourselves within our own consciousness. We are each a presence of awareness within infinite consciousness, able to create our human expression and everything we want to experience. When we follow our joy, we are aligning with life-enhancing energy and experiences. We are creating them with our vibratory presence.

How grateful we are about ourselves, or how much we may need to judge and criticize others or ourselves, determines the vibratory nature of our creations and the world we experience. A negative reaction to someone or something is based on fear and feeling threatened. Without our fear, a threat is meaningless. We are the ones who grant ourselves fear. There is no outside requirement. We are free to express our human person as we wish. All of our experiences are attracted to us as a result of our vibratory quality. When we have only good feelings, our life force radiates from us in ways that attract the people and experiences that we love and are grateful for.

Being grateful and loving in our present situation, whatever it may be, gives us a radiant presence that draws more love and gratitude into our lives. We are learning to align ourselves with the vibratory qualities of the consciousness that is our essence, our eternal presence of awareness with infinite creative power. We are the modulators of creative energetic patterns without limit within the consciousness that constantly creates everything.

The Challenge of Physical Death

When we believe and realize we are within a closed system, such as Newtonian physics or deductive logic, we can find proofs of what we know from within the system, but we cannot prove anything beyond it. So it is within the dualistic empirical world. We have created our senses to align our energy signature with its vibrations, and we believe and realize that this is our reality. From within the body, we cannot know what may lie beyond our senses, although we have created technology that operates with wave-lengths beyond our physical senses. Because these vibrations affect us, we can open ourselves to become emotionally-aware of them.

Because we recognize physical death and believe that it is real, we cannot prove from within the physical world, that we have life beyond time and space, although we have many accounts of witnesses of life beyond death. These give us the impression that we have life beyond the physical, but they do not prove it, because the experience is in a different plane or spectrum of consciousness.

For us to know our life beyond the body, we must open ourselves to a different kind of awareness. We can experience it in a way that is clear and real. It requires us to bring our personal state of being into transcending our limiting beliefs about ourselves and to feel ourselves living in the most wonderful conditions we can imagine. We can practice filling ourselves with gratitude and joy and love for everyone, as well as deep compassion for those who embrace darkness and attempt to prey upon us to sustain their life force.

If we want to transform our entire lives into what we truly love and can feel grateful for in every moment, we can intentionally engage with these life-enhancing energies. In order to know about life beyond the body, we must pay attention to our inner knowing and learn to trust it, because we are deeply connected with greatly-expanded consciousness. When we can understand everything through the consciousness of our heart, we can know the truth of our essence and the potential of our awareness. Our heart lives to enhance our lives. Our conscious life-force flows through the heart of our Being into our awareness to the extent that we allow with our limiting beliefs.

We are the directors of the limits of our awareness. At any time, we can change our direction by paying attention and aligning ourselves energetically with the life-enhancing energies of our heart. We can no longer identify with doubt or fear, because they keep our awareness locked into the dualistic world. They do not exist in the dimension of heart-consciousness. This is the dimension of love and abundance and intentional creation and expansion of all life. It brings greater stimulation and intensity of brightness and joy, as well as deep inner guidance and knowing beyond the body.

Dealing with Tyranny and Adversity

We are always being guided through the best situations for us, with a view of transcendence and alignment with the Consciousness of our Creator. When we realize that we have limitless awareness, everything that we gaze upon with our inner vision becomes real for us in our experience. We are learning to direct our attention to our inner vision by opening our awareness to the most life-enhancing inner guidance that we can pay attention to. Our attention and vibratory alignment are our energetic creative modulators of the quality of our presence. By our thoughts and emotions, we are constantly emitting photons with an electromagnetic radiance, attracting resonating energetic patterns to our attention, and that become our experiences.

If we learn to control and direct our attention to the vibratory spectrum that we love the most, we can transform our lives into encounters that we most want to pay attention to and experience. Outside of our own consciousness, there is no requirement for us to experience anything we do not resonate with. We live in an energetic band of vibrations that emanate from our heart and are shaped by beliefs about ourselves. By intentionally opening our awareness in gratitude to a realm beyond fear and doubt, we can feel and know the unconditional love that is the essence of our Being. Once we are aware of limitless consciousness, we can align our imagination and feelings with its vibrations. This alignment holds our attention in the vibratory range of creative joy and compassion.

When tyrannical events occur around us, we can understand that this is happening in our own consciousness, and we can change it within ourselves. Arising for us through the heart of our Being, our conscious life-force constantly guides us to safety and life-enhancement, if this is our desire. All possibilities are available to us in every moment. By being able to control and direct our attention constantly to what we can feel grateful for, we open our awareness to an expanded realm of realization beyond our ego-consciousness.

In this state of mind, having inner awareness of Self-Realization of our essence, we cannot be threatened with victimhood, because negative energy does not exist in our realization. Instead, we become masters of our situations, able to modulate the energetic patterns that we encounter with our own life-force flowing through our presence of awareness. This awareness can be directed to anything anywhere, creating the qualities of experiences that resonate with it. If physical danger is imminent, we are guided to safety in a different situation.

What appears to be happening in the world around us is all happening in our own consciousness. By directing our own perspective in opening our awareness to the light in every being, we exempt ourselves from being victims, and instead can act as masters of our human experiences. It's all about staying vibrationally high throughout our awareness and acting creatively. This means alignment with infinite love and enhancement of all life.

Transforming Our Consciousness

We can ask to be drawn more deeply into the consciousness of the Creator. Concurrently we can align our imagination and emotions to energetic patterns flowing with unconditional love in our essence. This is our vibratory frequency beyond time and space. It is the vibration of creative life-enhancement everywhere. We are present here, and in our essence, we are everywhere and have no physical presence, other than what we create for ourselves in every moment. In resonance with our thoughts and feelings about ourselves, we are constantly creating our lives and our bodies. By controlling and directing recognition and realization of our energetic vibrations and polarity, we can transform our bodies and experiences. All of this has been impossible for us to identify with, because our ego-consciousness cannot go there. Ego has a limited conscious awareness that cannot go beyond dualistic empiricism. This is our realized reality of time and space. It is a kind of hypnotic trance that allows for awakening at will by aligning with the consciousness of the Creator.

In our innate Self, we contribute to, and have access to, infinite consciousness. The quality of every energetic expression that we wish to observe, feel and recognize is creative. Without intuitive guidance, the ego creates chaos with fear and doubt. Ego has no true understanding of life. To gain this understanding, we must have a broader perspective and be willing to go beyond our comfort zone in faith and trust that we are sufficiently aware of our inner knowing. This state of knowing everything is available to us through our intentional energetic alignment with infinite love and joy.

These are the qualities of true creative power, in which there is absolute confidence in our inner knowing and feeling. We can recognize our expanding awareness beyond our former limiting beliefs about ourselves. We are playing roles in the human drama. Most of us keep to the script of our destiny, and then there are the ones who recognize some clues that point beyond the norm of awareness. We can recognize that everything is within our own consciousness, and we are the constant creators of the qualities of our roles through our energetic radiance.

As we transition to a new era, we pass into a life-enhancing world, while the world of entropy dissolves. Much of humanity is creating a path toward more duality and limitation, while threats increase and security melts down. This world is disappearing from our personal experience. We can withdraw our attention from it and instead devote our attention to the vibrations that speak from our heart. While the consciousness that we live within is filling with greater love, we can live every moment with compassionate understanding, as our creative ability grows more direct and powerful.

From Situational to Sensational Awareness

If we desire to expand our conscious awareness into mastery of this life, we must become the intentional directors of our mental and emotional processes. We must learn to direct our awareness beyond the body. We all do this when we daydream and at night, but we do not normally direct what's happening in our dreams. We allow our subconscious mind to sort through the impressions that it has received through the vibrations of our state of being during our waking hours. It then brings up to us scenarios that resonate with the energy signature we have created for ourselves by our imaginings, loves and fears.

In the trance that we realize as our empirical reality, we have learned to believe that we have no power to direct our lives. We live subject to our destiny, which we have designed prior to incarnation, as well as our karmic and dharmic creations, which our subconscious uses to bring us our experiences. Our dreams occur beyond ego-consciousness. The ego must go to sleep in order for us to dream, and the conscious mind must open to awareness beyond time and space in order to transcend the energetic spectrum of the human trance, opening to greater awareness.

As fractals of universal consciousness, we have abilities that we have kept ourselves unaware of, through our limiting beliefs about ourselves. From within the vibratory limits of the world we have recognized as real, we cannot open our awareness to the beyond without fear of losing everything we have accumulated mentally and emotionally, because we know nothing about what may be beyond our realized experience. In the beginning it is an adventure in consciousness, learning ways of living that we have believed are not sustainable or even possible.

Our challenge is that in the realm of the divine, we have only infinity without energetic limits. It is beyond mathematics and all operations of human thought. There is only knowing about everything, and we are tasked with creating more unique experiences, as much as we can imagine. This is an aspect of our being that operates constantly in alignment with how we feel about ourselves. Everything is within our consciousness, manifesting for us as we recognize it and realize its reality, as quantum physics experiments with sub-atomic entities have confirmed. Every quantum part of our bodies is in essence a conscious being in resonance with the vibratory presence that we create with our mental and emotional processes.

Living in infinite love and joy, while knowing everything that is happening wherever we focus our attention, gives us deep understanding and compassion for all who believe they are deficient in any way. Once we free ourselves from our self-imposed limiting beliefs, we can have awareness beyond our imaginings. While living in control of all aspects of our lives, we can access the full power of our creative life-force in alignment with the vibrations of the heart of our Being. This is our connection with our infinite Self. In our essence we are beyond time and space and need nothing that we have or are here. There is beauty and intense joy that we can experience as reality beyond the human trance.

Deepening Our Understanding and Expanding Our Awareness

Conscious awareness can be subdivided into unique bands of energy that exist as separate from other aspects of infinite consciousness. In this way we perceive and believe that we are localized in our physical body and are separate from other physical entities. In this vibratory spectrum we are unaware of the source of our being alive. We experience being alive, but how we get our aliveness is unknown. We know the expressions of our living processes, but we do not understand life itself. For this understanding we must make a leap in consciousness into our own non-localized Self-Awareness. Beyond ego-consciousness lies a greater reality that we are present in. In our present awareness, we can be aware of our presence beyond physicality. It is the essence of our life beyond space and time. This is where our understanding of life comes from.

Since words lie within the limits of time and space, we cannot describe greater conscious realization, but we can begin with imagining greater enhancement of life and searching in our inner knowing. If we are too entranced in the dualistic empirical world, we may need a traumatic experience to awaken us. Such experiences are always symbolic, prompting a movement through terror, loss, trauma, grief and sorrow. We can understand that all of this is based on fear of ultimate termination. We believe in our demise, and thus can be threatened and victimized. If we desire to resolve this limiting belief, we must examine it closely to find its origin. Here we find that fear of our demise is self-created, based on not knowing what may lie beyond the physical. We could at least be neutral, since we do not know.

Being neutral allows for possible opening of awareness to our expansive essence. It is awareness of the energy of the heart of our Being, which we know intuitively. When we align our imagination and emotions to its resonance, we can connect with our deeper knowing in gratitude, love and joy. In this state of Being, all fear and doubt dissipates, because it no longer has a basis, and is not believable.

It's a challenge to face everything that life brings to us with love, compassion and unattachment, but in our inner knowing we can be aware of our eternal presence guiding us through every experience. We must want to have this awareness and practice having it, perhaps at first in our imagination and by paying attention to what arises within. We can ask for inspiration from our friends who live in a non-localized realization. If we want to live in a world of love and beauty, we must align our mental and emotional processes to resonate with the vibratory spectrum of this world.

As we open our awareness into closer alignment with the heart of our Being, we align with the consciousness of our Creator. Everything that is not in resonance with this disappears out of our experience. From the perspective of ego-consciousness, this can be traumatic and sorrowful, because the ego does not know what is happening in the realm of cause. Living in the consciousness of our Creator becomes possible and opens our awareness to the life we truly want to experience.

www.ingramcontent.com/pod-product-compliance
Lightning Source LLC
Chambersburg PA
CBHW070945180426
43194CB00040B/946